WASP of the Ferry Command

WASP OF THE FERRY COMMAND

Women Pilots, Uncommon Deeds

Sarah Byrn Rickman

Foreword by Deborah G. Douglas

UNT PRESS
Denton, Texas

Printed in the United States of America.

10 9 8 7 6 5 4 3 2

Permissions:
University of North Texas Press
1155 Union Circle #311336
Denton, TX 76203-5017

The paper used in this book meets the minimum requirements of the American National Standard for Permanence of Paper for Printed Library Materials, z39.48.1984. Binding materials have been chosen for durability.

Library of Congress Cataloging-in-Publication Data

Names: Rickman, Sarah Byrn, author.

Title: WASP of the Ferry Command : women pilots, uncommon deeds / Sarah Byrn Rickman ; foreword by Deborah G. Douglas.

Description: Denton, Texas : University of North Texas Press, [2016] |

Includes bibliographical references and index.

Identifiers: LCCN 2015046031| ISBN 9781574416770 (paper : alk. paper) | ISBN 9781574416428 (ebook)

Subjects: LCSH: Women Airforce Service Pilots (U.S.)--History. | United States. Army Air Forces. Air Transport Command. Ferrying Division--History. | World War, 1939-1945--Participation, Female | World War, 1939-1945--Aerial operations, American. | World War, 1939-1945--Transportation--United States. | United States. Army Air Forces--Transportation. | Women air pilots--United States--History--20th century. | Airplanes--United States--Ferrying--History--20th century.

Classification: LCC D790.5.R53 2016 | DDC 940.54/4973--dc23

LC record available at http://lccn.loc.gov/2015046031

The electronic edition of this book was made possible by the support of the Vick Family Foundation.

The WASP, the Women Airforce Service Pilots of World War II, did not evolve out of a void. During World War I at least two American pioneer aviatrixes, Marjorie Stinson and Ruth Law, taught World War I military cadets to fly.[1] *By the mid-1930s, led by Amelia Earhart, Louise Thaden, Phoebe Omlie, Ruth Nichols, Helen Richey, and others, women were leaving their mark on the aviation world. In the summer of 1942, the Ferrying Division, Air Transport Command, U.S. Army Air Forces offered women pilots the opportunity to fly for their country. This book is dedicated to the 303 women—known ultimately as WASP —who ferried military aircraft for the FD/ATC from October 1942 to December 1944 to release male pilots for combat and other aviation duties during World War II.*

Table of Contents

List of Photos

Foreword

by Deborah G. Douglas

Upon learning that England was recruiting women pilots in 1942, Eleanor Roosevelt wrote in her famous "My Day" newspaper column that American women pilots were a "weapon waiting to be used."[1] Days after her column appeared, two programs got underway to do just that: allow women to fly military airplanes. One recruited extremely experienced women pilots and formed a small squadron that was immediately integrated into the operations of the Army's Air Transport Command. The other recruited licensed pilots but with much less experience and put them through the Army Air Forces Training Command's typical cadet flying school program. In 1943, the two programs were combined to form the Women Airforce Service Pilots or WASP. In December 1944, twenty-seven months later, with the war now turning in favor of the Allies and a reduced need for pilots, the WASP program was deactivated.

General William H. Tunner, one of the military officers who had done the most to fulfill Eleanor Roosevelt's vision, wrote in a letter of commendation: "The WASPs of the Ferrying Division leave behind a truly impressive record, one with which I am sincerely proud to have been associated. They have accomplished far more than the safe and efficient delivery of hundreds of vitally needed aircraft; they have proved beyond all doubt that in times of national emergency, America can give its women the most challenging assignments with complete confidence."[2] General Henry H. "Hap" Arnold, Commanding General of the Army Air Forces, put it this way: "It is on the record that women can fly as well as men."[3] Eleanor Roosevelt had been right: American women—whether flying the planes, building them, or just spotting them—had proven an extraordinary national asset in a time of great need.

Sarah Rickman's new book is all about the details of those twenty-seven months. Think about this. In just over two years, 1102 women profoundly altered the American social landscape. To my knowledge, not a single one signed on with this in mind. These were women who against many odds had fallen in love with airplanes and by chance were *exactly* the right age, health and social circumstances to be invited to take part in a radical experiment that allowed women to pilot military aircraft. As Cornelia Fort put it in an article for *Woman's Home Companion* in 1943: "I, for one, am profoundly grateful that my one talent, my only knowledge, flying, happens to be of use to my country when it is needed. That's all the luck I ever hope to have."[4]

So what happened in those twenty-seven months? Rickman has chosen to tell this story through the eyes of the 303 women who ended up serving the Ferrying Division of the Air Transport Command plus the women and men who led them. The WASP did many things beside ferry aircraft (from towing targets to serving as test pilots) but this group represents the heart and soul of the program. It also became the chief battleground for the competing visions (and egos) of Nancy Love and Jacqueline Cochran, the original leaders of the two programs initiated in 1942.

Rickman immerses the reader into every aspect of the world of the WASP of the Ferry Command. There are excerpts from original documents and interviews and a close reading of the secondary sources. She scrutinizes these women from every angle. It is the kind of detailed analysis that will delight aficionados of the WASP but will also reward historians interested in better understanding the complex power struggle that culminated in the abrupt termination of the WASP program. It clearly did not have to end the way it did.

"Yet what if it hadn't?" That is the question I found myself musing on for the first time. What if the WASP had been militarized in 1944? Obviously, the women would have received key benefits at the war's end but would they have received the Congressional Gold Medal? Would there have been a special float in the Rose Bowl Parade? Would women be flying military aircraft or serving in combat today? In other words, did the fight over militarization ultimately lead to greater recognition and have far greater impact than might otherwise have been the case?

Thorsten Veblen said, "The outcome of any serious research can only be to make two questions grow where only one grew before." Sarah Rickman's new book is filled with serious research and therefore it will stimulate a myriad of new questions. But it is also a loving homage to her heroines and all readers will enjoy her vivid descriptions and distinctive perspective on the story of these pioneering women and their "uncommon deeds."

Deborah G. Douglas

MIT Museum

Author of *American Women and Flight since 1940*[5]

Acknowledgments

My most grateful *"thank you"* to the National Aviation Hall of Fame, Dayton, Ohio. In 2009, the NAHF bestowed on me the Combs Gates Award. This award is given annually for creative projects that "reflect an emphasis on the individual pioneers —the people—who defined America's aerospace horizons."[1] My two earlier books, *The Originals: The Women's Auxiliary Ferrying Squadron of WWII* (2001) and *Nancy Love and the WASP Ferry Pilots of WWII* (2008), and a draft and promise of this volume were what brought the award my way. It made possible the ensuing six years of work it took to produce this book—the final installment of my "WASP of the Ferry Command" trilogy.

My thanks to Iris Cummings Critchell, WASP Class 43-2, and to Barbara "B.J." Erickson London, WAFS, my mentors during the research and writing of this book. Sadly, before it was completed, "B.J." took her final flight—July 7, 2013. Or as aviators say, "she flew west." I miss her.

Sarah Byrn Rickman

October 3, 2015

WASP of the Ferry Command

Introduction

Eleven-hundred-two American women flew U.S. Army aircraft in World War II. Those women, the WASP (Women Airforce Service Pilots), were WWII's best-kept secret.

The accepted story is that the WASP originated with Jacqueline Cochran, the cosmetics company founder who made her name racing airplanes in the 1930s. Cochran's contribution is only half the story. The full story of the WASP is much bigger and far more complex. The desire to tell this larger story led me to research and write *WASP of the Ferry Command.*

Chapter One relates the pre-and-post-Pearl Harbor events that led to the spring 1942 establishment of the Air Transport Command and Ferrying Division, U.S. Army Air Forces. Chapters Two and Three introduce the WAFS—the Women's Auxiliary Ferrying Squadron—the first twenty-eight experienced women pilots who were recruited by the Ferrying Division to move liaison and primary trainer aircraft from the factories to the flight training schools. The group was formally announced September 10, 1942. Their leader was Nancy Harkness Love, another accomplished woman pilot of the 1930s.

Chapter 4 introduces the Army flight training school for women established through Jackie Cochran's advocacy and by order of Commanding General of the Army Air Forces H. H. "Hap" Arnold. It opened November 1942. Chapters Five through Eight tell about the women graduates of the school's first six classes. Those women were destined to fly for the Ferrying Division. Chapters Nine through Thirteen describe the Ferrying

Division's formation of four women's squadrons, all located near aircraft factories building the primary or basic trainers the women would be ferrying. The original WAFS and the graduates of those six classes were assigned to those four squadrons. After the sixth class, few women graduates were sent to the Ferrying Division, more popularly nicknamed the Ferry Command.

The remaining chapters relate, as they unfold, the events that mark the ongoing power struggle between the two women leaders, their respective military supporters, and their divergent philosophies of how to deal with women pilots in the military—all of this uncharted territory.

How the Love-Cochran rivalry plays out against the backdrop of the ever-changing war scenario provides a compelling, but vexing story as needs, priorities, and missions change.

WASP of the Ferry Command is laced with stories of ferrying flights gleaned from oral histories housed in the WASP Archives and from a few author one-on-one interviews with some of Nancy Love's women ferry pilots. The work of ferry pilots was said to consist of "hours of boredom interspersed with moments of sheer terror!" There were humorous moments as well.

A very important note: The use of oral histories created a problem. In oral histories, the narrative tends to ramble off subject and be filled with interruptions only to return to a story a paragraph or a page later. To "cut" the story and "paste" it on the page of the manuscript so that the reader gets its full impact presented a technical literary problem. Good writing demands clarity of thought and engaging material. And, the actual words spoken by an individual need to be put in quotation marks and then attributed to the speaker.

Attribution is no problem, but word-for-word accuracy is when it comes in conflict with literary flow and the telling of a story. The solution has been to offer these stories—the actual words of the women themselves—to you in a clear sentence and paragraph form *set in* **italic**

type to set them off. If you read the oral history, you will have to sleuth out the entire thought from among many pieces of information—as I did. But this way, the story is related in a way that it is a complete thought. So when you read these italicized stories in several of the chapters, you ARE reading a WASP's words, but not necessarily in the exact order that she related the story to the interviewer.

In 1999, Original WAFS Nancy Batson Crews, challenged me to write the little-known story of Nancy Love and the WAFS. That challenge produced *The Originals* in 2001 and the Nancy Love biography—*Nancy Love and the WASP Ferry Pilots of World War II*—in 2008.

Iris Cummings Critchell, a graduate of the school's second class (43-2) and who became one of Nancy Love's most versatile ferry pilots, urged me to complete the trilogy by writing about the women from the Army flight training school who went on to serve as pilots for Nancy Love and the Ferrying Division. Just as Nancy Crews made the first two books possible, Iris—by painstakingly sharing with me the story of the women of the "Ferry Command" as she knows it—made possible the book you hold in your hands.

Getting to know the WASP gave me the needed incentive to fulfill a lifelong desire to learn to fly. To emulate them, I took flight instruction in a 65-horsepower Aeronca Champ, a two-seater taildragger very much like the J-3 Cubs the WASP learned in back in the early 1940s. Flying off a grass strip at Red Stewart Airfield in Waynesville, Ohio, I earned my Sport Pilot's certificate July 1, 2011. Thank you Emerson Stewart III and the late John Lane. And now, the book, too, is finished!

What a journey it has been!

Photo 1. Original WAFS Barbara "B.J." Erickson and Evelyn Sharp, Long Beach ferry pilots, were the second and third women to check out in the P-51/A-36, June 1943. Nancy Love was the first.

Photo courtesy of the Iris Cummings Critchell Personal Special Collection

Part I

War Declared, Pilots Needed

September 1939–September 1942

When the Germans invaded Poland on September 1, 1939, the world was plunged into what would become the second worldwide war of the twentieth century. Though the United States didn't enter the fray as a combatant until December 8, 1941, following the December 7 bombing of Pearl Harbor, the country nevertheless was affected by everything that was to come.

In 1938, President Franklin D. Roosevelt approved the Civilian Pilot Training Program (CPTP), "the first full-scale federally funded aviation education program ... to provide a pool of young civilian pilots who would be available for military service if war came."[1] CPTP and its successor the War Training Service (WTS) ultimately trained 400,000 young men to fly between 1939 and 1944, most of whom ended up flying military aircraft in World War II.[2] In addition, CPTP admitted women on a one-to-ten ratio to men from 1939 through June 1941. That increased the number of women pilots in the United States by roughly 2,325: from 675 in 1939 to almost three thousand by July 1, 1941.[3] Because of two events that occurred in September 1942, many of those young women went on to fly military aircraft in WWII.

On September 10, 1942, Gen. Harold L. George, commander of the Air Transport Command (ATC), U.S. Army Air Forces, introduced eminent woman pilot Nancy Harkness Love as the director of his newly formed women's squadron. The term Women's Auxiliary Ferrying Squadron was not used in the announcement, but the original squadron of women pilots, which numbered twenty-eight, would ever after be known by the acronym WAFS, and also as "The Originals." They were hired to ferry small military trainer and liaison aircraft for the ATC's Ferrying Division. Ferry pilots move aircraft—not people and goods—from one place to another.

On September 15, 1942, Commanding General of the USAAF Henry H. "Hap" Arnold announced that celebrated woman pilot Jacqueline Cochran would form and lead a second women's program—the Women's Flying Training Detachment (WFTD)—an Army program to train more women pilots to ferry aircraft for the Ferrying Division's women's squadron.

Chapter 1

The Beginnings

The world in the late 1930s waited, poised on the brink of war.

After World War I, U.S. Army Air Service Brigadier General Billy Mitchell predicted that the next war would be won in the air. This heresy got him court-martialed.[1] But not all of the Army brass thought Mitchell was wrong. One such man was H.H. "Hap" Arnold, who went on to serve as Commander of the U.S. Army Air Forces during World War II. Another was William H. Tunner, a member of the West Point graduating class of 1928.

"The Air Corps was considered the lunatic fringe," General Tunner wrote in his memoir. But he opted for it anyway. Five flights in five different airplanes in one week's time during his senior year had shown him that "man could fly."[2]

In 1938, with the threats of Nazi Germany and Imperial Japan on the rise, Captain Tunner saw the growing need for a corps of reserve pilots. He began to recruit local pilots in Memphis, where he was stationed, and built a sizeable group of Reserve flying officers.[3]

Women also recognized the coming need for more pilots. Cecil "Teddy" (Mrs. Theodore) Kenyon, a founder of the Ninety-Nines—the international organization for licensed women pilots[4]—suggested in 1936 that women could assist in wartime by ferrying planes.[5] Betty Huyler Gillies, another Ninety-Nines founder and the organization's president from 1939 to 1941, advised the membership to sharpen their skills and build their time. "Be ready to fly for your country if she needs you."

In 1939, Jacqueline Cochran—founder of Jacqueline Cochran Cosmetics, a record-setting racing pilot, and a Ninety-Nine—wrote to Eleanor Roosevelt to suggest a plan to use American women pilots in a national emergency.[6] Women could fly ambulance planes, courier planes, and commercial and transport planes, "thereby releasing male pilots for combat duty. ... We have about 650 licensed women pilots in this country. Most of them ... could be of great use a few months hence if properly trained and organized."[7]

In 1940, another well-known aviator and Ninety-Nine, Nancy Harkness Love, wrote to Lt. Col. Robert Olds, in the Plans Division of the Office of the Chief of the Air Corps, about the possibility of women ferrying airplanes for the U.S. Army. Mrs. Love had gained experience delivering airplanes that she and her husband, Robert M. Love, sold to their customers at InterCity Aviation in Boston. Olds asked for a list of women pilots holding commercial ratings. Using Aero Chamber of Commerce lists, Nancy Love gave him forty-nine names. "I really think this list is up to handling pretty complicated stuff."[8]

That summer, Olds considered using "approximately 100 women pilots as co-pilots in transport squadrons and for ferrying single-engine aircraft thereby releasing a number of [male] pilots" for other duty. General Arnold turned down the proposal.[9]

The United States was still a neutral country, but civilian crews employed by the Atlantic Ferrying Organization (Atfero) were moving aircraft from Canada across the North Atlantic for the British.[10] Then in March 1941, Lend-Lease—through which the United States supplied the United Kingdom, the Soviet Union, the Free French, China, and other Allied nations with materiel between 1941 and 1945—went into effect.[11] The Army Air Corps began to plan how the United States could deliver aircraft to England and, in doing so, give its ferry pilots on-the-job training and experience.[12]

On June 20, 1941, the Army Air Corps ceased to exist, replaced by the U.S. Army Air Forces (USAAF) and Colonel Olds was assigned to set up its Ferrying Command. Major William H. Tunner was the second man Olds brought on board.[13] Immediately, Olds began to establish routes by which the new Ferrying Command could move aircraft overseas.

"Our knowledge of the world was expanding to be followed by the continual expansion and increase of our actual flying operations," Tunner wrote.[14] His job was to find ferry pilots to fly planes to various destinations. He tried to fill the ranks with National Guard and Reserve pilots, but there weren't enough, so he asked for the loan of pilots from tactical (planned/strategic) units.[15] This practice would continue until fall 1943.

The United States expected to build 12,000 training-type aircraft by December 1942. Two hundred pilots were needed to deliver those aircraft. Fifty women had at least five hundred hours; thirty of them had more than a thousand. Using those women would solve one-quarter of the Ferrying Command's needs, but Arnold again said no.[16]

Four months later, Pearl Harbor changed everything.

On December 7, 1941, Japanese aircraft attacked the United States' military installations on the Hawaiian island of Oahu, destroying much of the Navy's Pacific Fleet anchored in Pearl Harbor. On December 8, the U.S. declared war on its attackers, the Japanese, and also Japan's ally, Nazi Germany. Everybody who was anybody when it came to moving men and supplies by aircraft gathered in Washington, D.C., on December 14, 1941, to get ready for a war that was already upon them.

Logistics was the problem, wrote Lt. Col. Oliver LaFarge, Air Transport Command historian. "America and her allies were weak in men, ships, aircraft, and munitions of all kinds."[17]

As they were meeting, General Douglas MacArthur's forces in the Philippines were under a fierce Japanese attack. Though no supply route yet existed, the aircraft movement brain trust sent the few heavy transports the U.S. could muster to supply MacArthur. Not one plane had reached its destination by the time MacArthur was forced to retreat to the Bataan peninsula in late December. Some planes got to the Dutch East Indies in time to load their munitions on submarines; others never got beyond India. Three reached Australia.[18] There was no good news from the Pacific.

"We forget how feeble we were in the beginning of 1942, and how black the picture was," LaFarge wrote. The lack of long-range transport was perhaps the worst problem, but already the Ferrying Command was using its "handful of specialists who had learned how to fly oceans and unexplored continents" to train more ferry pilots to do the same.[19]

"They were charged with creating and operating the ferry routes across the North Atlantic, the South Atlantic, and Africa and across India to Burma and China," says Iris Cummings Critchell, aviation historian and one of the early WASP ferry pilots. And, she pointed out, if the Ferrying Command and its successor the Ferrying Division hadn't been effective, many other commands would not have succeeded in their missions because the Ferry Command was the supply line.

"That is my view of the historical record of some remarkable planning by those early military aviators who had been active in the previous 20 years in aviation and came back to serve in World War II as the leaders. They had to have an appreciation of not only the limitations of the human beings and the need to train them, but also the capabilities of the human being to do a full job. These men understood what it took to be a pilot and they placed confidence in the new ferry pilots who were challenged by the times and the circumstances to do their best. And they did."[20]

***

Immediately after Pearl Harbor, Tunner's "borrowed" ferry pilots were called back to their units. In January 1942, the total strength of the Ferrying Command stood at a mere 934 men: 386 pilot officers and 548 enlisted, men who had come from the National Guard or the Reserves.[21]

On January 6, President Roosevelt delivered his annual State of the Union address to Congress. He promised that, in 1942, the United States would produce 60,000 airplanes. In 1943, it would produce 125,000—100,000 of them combat aircraft.[22]

All those new aircraft would have to be ferried and Tunner and Olds did not have nearly enough pilots to do the job. Finding ferry pilots became a priority. Olds, who had been promoted to Brigadier General, decided to resurrect Nancy Love's idea. He was ready to hire women pilots, right then, as civilians—the same way he was hiring male pilots where he could find them. But Jacqueline Cochran—with the backing of General Arnold—had made an unbreakable commitment to the British to supply them with twenty-five of America's best women pilots for ferrying duties. She immediately wrote the following note to Hap Arnold dated January 18, 1942.[23]

> General Olds has informed me that he is planning on hiring women pilots for his Ferrying Command almost at once. His plan, as outlined to me, is not only bad in my opinion from the organizational standpoint, and contrary to what you told me yesterday but is in direct conflict, in fact, with the plans of a women's unit for England. In addition, it would wash me out of the supervision of the women flyers here rather than the contrary as we contemplated.

Arnold was aware of Cochran's political clout with the Roosevelts (her wealthy husband, Floyd Odlum, was a financial and political backer). Arnold wrote to Olds and told him: "You will make no plans or re-open negotiations for hiring women pilots until Miss Jacqueline Cochran has completed her present agreement with the British authorities and has returned to the United States."[24]

Reluctantly, Olds put his idea on the back burner.

***

At this point, the idea of the AAF's Ferrying Command using women pilots appeared to be dead.

That, however, was not the case.

Yet to come was the fortuitous, and consequently historic, meeting of minds and purpose of four high-ranking Ferrying Command officers and one civilian woman who were about to change everything.

***

All airfields within fifty miles of the U.S. coastline were shut down after the attack on Pearl Harbor. Nancy and Bob Love's InterCity Aviation in Boston was one of the operations affected. Concurrently, Bob Love, an Air Corps Reserve officer, was called to active duty in Washington.

On January 5, 1942, Maj. Robert H. Baker arrived at Logan Field near Baltimore, Maryland. Baker had been a flying officer in World War I. He entered the flying service in April 1917, and became a flight instructor at Ellington Field, Texas, and at Wright Field, Dayton, Ohio. He served as officer in charge of flying at Post Field, Fort Sill, Oklahoma, and left the service in June 1919. In June 1927, he joined the Arkansas National Guard where he served as Captain and Operations Officer of the 154th Observation Squadron. Baker was called up in June 1941, commissioned a major, and sent to Detroit with the newly formed Ferrying Command. As part of President Roosevelt's Lend Lease program, bombers were being ferried from there to England.[25]

In Baltimore, Major Baker's orders were to set up the Northeast Sector of the Domestic Wing of General Olds's Ferrying Command.[26] The Domestic Wing was, as it suggests, responsible for flights within the United States. The Foreign Wing handled overseas flights.

Olds recommended that Major Baker hire Nancy Love, who had joined her husband, Bob, in Washington. Through her work with her husband's business, she was experienced both in ferrying and running air operations. Nancy went to work for Baker in the Operations Office of the Northeast Sector on March 11, 1942.[27] Her job included mapping ferry flights and routes, learning military procedures, and helping find sources for pilots.

That same month, Bob Olds, whose health was problematic, collapsed at his desk. Hap Arnold brought in Col. Harold L. George, a recognized strategic bombing expert, as a temporary replacement. In the 1930s, George had helped shape America's bomber strategy for the war by assisting Air War Plans Division with the development of a complete aircraft production and bombing strategy.[28] But Arnold knew what he wanted and he thought George was the man to head it up.

General Arnold's major concern in the spring of 1942 was the building of a globe-encircling military airline. "The attack on Pearl Harbor brought us renewed pleas from all parts of the world for personnel, airplanes, engines, parts, and supplies," Arnold wrote in his memoir *Global Mission*. "Airplanes, as weapons of war, must not be held on the ground. ...We had to keep them in the air and keep them moving. This required supplies and replacements in a constant stream."[29]

When Arnold learned that Olds's recovery would be slow, he relieved him of his command and asked Colonel George to take over the Ferrying Command. George was expecting an overseas command, not the stateside job of running what amounted to a major airline. He told Arnold he knew nothing about moving men and supplies by air. Nor did he know anyone in the Ferrying Command who had the expertise to create the worldwide air transport system the Army needed to accomplish the daunting task.

Did George know C.R. Smith? Arnold asked.

George knew of him—the president of American Airlines—but didn't know him personally.

Arnold knew Smith, called him, and set up a meeting in Washington. The chemistry was right. George and Smith could work together. Arnold made George a brigadier general and offered to make Smith a colonel and name him George's chief of staff.[30] Forty-two-year-old Smith agreed. He would serve as Harold George's deputy commander and eventually became a general himself.[31]

"Once I had George and Smith running our military air lines, I didn't have to worry about that problem any longer," Arnold wrote. "No matter what mission I gave them, I could count on its being carried out one-hundred percent. The two officers complemented each other in ability, experience, and judgment—they made a perfect team."[32]

Smith quickly solved any staffing problems George might have had. He had in his possession the names of several airline executives and, with George's nod of approval, he began to recruit these men to do the job the command needed done.[33] Bob Love was one of them.

Major Baker was confronted with two problems at Logan Airport. His pilots were responsible for ferrying the twin-engine B-26 bombers built at the Martin plant in Baltimore, but the runways at Logan—3,400, 3,000, and 2,000 feet long, respectively—were too short for safe operation of the big bombers. The other concern was the constant B-26 traffic on the runways and parking areas interfered with private aircraft traffic, which created problems for Baltimore's civic leaders.[34]

New Castle Army Air Base was under construction outside nearby Wilmington, Delaware. Already, the Army had built two new 7,000-foot runways there. Baker liked what he saw. The Northeast Sector of the Domestic Wing, Ferrying Command's move to New Castle AAB was announced April 18, and was completed on May 29, 1942.

Meanwhile, Nancy Love, from her key role with Baker's Baltimore operation, had been observing, first hand, the ever-increasing production of military aircraft. Baker's pilots also were responsible for ferrying primary trainers, PT-19s, built at the Fairchild plant in Hagerstown, Maryland, and PT-19 production was humming.

To Nancy, it was clear that there were not enough qualified male pilots eligible to move all of those planes. Again, she proposed, this time to her superiors in the Northeast Sector, "that a group of highly-qualified women pilots be recruited to aid the Domestic Wing in its ferrying activities and as a test group to determine the feasibility of using women pilots."[35]

Though staff of the Domestic Wing opposed the idea, General George and the Ferrying Command supported it.[36]

In the three months that lay between March 11, when Nancy started working in Operations, and June 11, conversations among General George, Colonel Tunner, Major Baker, Major Love, and Nancy Love, must have been not only lively but intense. June 11 is the date of the earliest written mention of General George's decision to use women pilots.

In a June 11, 1942, memo that George wrote to the Assistant Chief of Air Staff, Personnel, AAF Headquarters, he stated that, "It is desired to use commissioned officers of the Women's Auxiliary Army Corps [WAAC] for the ferrying of airplanes in this Command, to replace and supplement male pilots in so far as qualified women may be available."[37] The idea of placing women ferry pilots in the WAAC was under discussion.

Working with the WAAC did not materialize, but by then the genie-of-an-idea—women ferry pilots—was out of the bottle.

Effective June 20, 1942, General George renamed the Ferrying Command the Air Transport Command (ATC) and the Domestic Wing became the Ferrying Division. Baker had been promoted to Lieutenant Colonel when he moved from Baltimore to New Castle. He was promoted to full Colonel, July 17, 1942.[38] His Northeast Sector of the Domestic Wing command became the 2nd Ferrying Group, one of seven ATC ferrying groups strategically located around the country.[39]

The stage was set.

Chapter 2

A Coup at the Water Cooler

The story, as Gen. William H. Tunner tells it in his memoir, unfolded like this:

Colonel Tunner was chatting with Major Love at the water cooler early one morning when Love mentioned that he hoped the fog had lifted enough for his wife to get off the ground and fly to work in Baltimore.

"Your wife flies?" Tunner asked.

Love answered in the affirmative.

"Good Lord," Tunner recalls saying, "I'm combing the woods for pilots, and here's one right under my nose. Are there many more women like your wife?"

Love, according to Tunner, replied, "Why don't you ask her?"[1]

Yes, Nancy Love told Tunner, she knew of nearly a hundred experienced women pilots.

Photo 2. Gen. William H. Tunner, Commander of the Ferrying Division, Air Transport Command.

Photo courtesy of the National Museum of the United States Air Force.

When Bob Love told Bill Tunner that his wife, Nancy, flew, he didn't elaborate. He let his wife speak for herself. Nancy Love had 1,100 hours' flying time and held her private and commercial licenses and her instrument rating. She had worked for the Bureau of Air Commerce's Airmarking program in 1935 and then as a test pilot for the experimental Gwinn Aircar and the Hammond Y aircraft in 1937–38. From 1938 through the end of 1941, she worked as a sales and demonstration pilot for InterCity Aviation in Boston, the company founded and run by her husband, Robert M. Love. One of her biggest responsibilities was ferrying aircraft to customers around the country. By spring of 1942, Nancy was working for Major Baker and the Ferrying Command in Baltimore. Her savvy when it came to aircraft operations, plus her team-player approach, had made her invaluable to Baker and his staff. Now it was Tunner's turn to meet the accomplished Mrs. Love. She would be lifelong friends with both of these men.[2]

In Nancy Love, Tunner had found a partial solution to his lack of ferry pilots. He hired her to find and recruit women pilots. First, he suggested commissioning women pilots as officers in the recently formed (May 1942) Women's Army Auxiliary Corps (WAAC). However, technical difficulties in the WAAC legislation currently before Congress stood in the way. Waiting for an amendment would take time, so the ATC decided to pursue hiring women pilots as civilians.[3]

Already Tunner was recruiting any civilian male pilots he could find. He hired them through the Civil Service Bureau on a three-month trial basis at $300 a month and a per diem of $6 when they were away from base on a ferrying assignment. If they performed satisfactorily, after ninety days they were commissioned in the Armed Forces and assigned to the Ferrying Division, Air Transport Command (FD/ATC). Tunner opted to do the same thing with the women. "They wanted to determine the suitability of using women pilots in the delivery of military aircraft," said veteran woman pilot Betty Huyler Gillies.[4] Betty was the first woman to join Nancy Love's squadron of women pilots.

Photo 3. Nancy Harkness Love, Executive for WASP of the Ferrying Division.

Courtesy WASP Archive, Texas Woman's University, Denton, TX.

The women's unit would consist of no more than 50 pilots and would be assigned to the 2nd Ferrying Group in Wilmington. The Civil Service

Bureau agreed to pay women $250 a month on a three-month trial. The initial intent was to give the women who qualified a commission in the Armed Forces, the same as they gave male pilots, but General Arnold was not in favor of militarization at that point.

Gillies pointed out that "ATC strongly supported that idea, wanting to handle the women exactly as they were handling the men." That was to be ATC's stance throughout the existence of the women pilot program —that men and women should be handled in the same manner.

Nancy Love drafted the plan for the proposed women's squadron —the original was in her handwriting—and on June 18, Tunner sent General George their proposal for utilizing women pilots.[5] General George forwarded the proposal to General Arnold.[6]

The proposed qualifications for women pilots were: 500 hours, commercial license, between the ages of twenty-one and thirty-five, and a high school diploma. Added later, on Nancy Love's suggestion: a 200-horsepower rating, two letters of recommendation, and U.S. citizenship. The salary of $250 a month was $50 less than male civilian ferry pilots received, because the women would be ferrying only light, less complicated aircraft. However, the women would receive the same $6 per diem when out on a ferrying trip.[7]

Tunner was so sure that their plan would be approved, the same day he sent the memo (June 18), he went to NCAAB to talk to Colonel Baker. The base was under construction. Tunner hoped to line up living quarters and mess privileges for the women pilots. Colonel Baker was most cooperative, finding living quarters for the women and offering them privileges in the new officers' mess.

Nancy Love had been working for Baker for several months by that time. In his June 20 memo to Colonel Tunner, Baker suggested that she be commissioned a First Lieutenant and assigned as Operations Officer for the women pilots. "I am confident that she can handle the assignment very efficiently; in fact, I do not know of anyone who has as good qualifications as she has for the assignment."[8]

Tunner chose NCAAB because the women could be part of the 2nd Ferrying Group. It was close to Hagerstown, Maryland, home of the Fairchild Aviation Corporation that built the primary trainers (PTs) he needed the women to ferry. He notified General George that things at Wilmington were a go and, on June 30, Colonel Baker suggested that Nancy Love "come to Wilmington immediately and be employed in her former status until commissioned and until the other arrangements were completed." In a July 5 memo, Colonel Tunner concurred.[9]

Nancy Love was on duty at the 2nd Ferrying Group as of July 13. That day she approved a memorandum from Colonel Baker regarding the training course for the women pilots as well as a standardized uniform to be worn by the women while on flying duty or on Post.[10]

On July 18, General George informed General Arnold that commissioning the women pilots in the WAACs would not work. No provisions existed for commissioning women flying officers or providing flight pay.[11] George also sent Arnold a memo proposing that the women pilots be hired as Civil Service employees for the purpose of "determining the suitability of utilizing women pilots in the delivery of military aircraft." "If you approve this," George said, "we will go ahead at once with the organization although you intimated, when we discussed this about three weeks ago, that you wanted to mention the matter to the President thinking that he would like to announce the formation of a Women's Pilot Corps for use in the domestic ferrying of aircraft."[12]

Rather than giving his approval at that point, General Arnold asked for more up-to-date information on women pilots through the Civil Aeronautics Administration (CAA) and the Civil Air Patrol (CAP). But he did add the following second paragraph to that memo dated July 20:[13]

> After exhausting that supply, then reopen this proposition. Give me an outline of the number of women pilots you expect to get and require to carry out your mission.

Part of Nancy Love's job included writing the women pilots' training syllabus, which she and Lt. Joe Tracy, one of Colonel Baker's top flight instructors, did in thirty days.[14] She also was appointed an unofficial member of the three-man board that would review the women applicants.[15] During August, no further word came from General Arnold but Nancy, Colonel Baker, and Colonel Tunner continued their planning.

On September 1, Mrs. Roosevelt wrote the following in her syndicated column "My Day":[16]

> It seems to me that in the civil air patrol and in our own ferry command women, if they can pass the tests imposed upon our men, should have an equal opportunity for noncombat service. ... Women pilots ... are a weapon waiting to be used.

On September 3, General George again recommended to General Arnold that a group of women pilots be organized in order to determine their suitability for delivering military aircraft. Again, he outlined the standards they must meet and explained that they would be hired through Civil Service. George proposed that a woman who would serve directly under the Group Commanding Officer handle the administration duties. Nancy Love's name was not mentioned, but she met all the requirements.[17]

On September 5, General Arnold directed that immediate action be taken and that the recruiting of women pilots begin within twenty-four hours.[18] Mrs. Roosevelt's column, Arnold insisted, had nothing to do with his decision. "The development," he said, "was all part of a regular plan."[19]

Nancy Love was appointed director of the woman pilot group.

General George told Nancy he had the go-ahead from General Arnold and to ready telegrams for her top candidates. With George's approval, she sent out eighty-three telegrams to prospective women ferry pilots on September 5, 1942.[20]

General Arnold was to make the public announcement on the morning of September 10. When Nancy Love and General George arrived at his office that morning, Arnold unexpectedly had been called out of Washington. Secretary of War Henry Stimson made the announcement instead and the swearing-in took place in his office. Then Secretary Stimson introduced Mrs. Love to the press.[21]

The afternoon of September 10, 1942, following the announcement by the Secretary of War, Nancy Love returned to Wilmington to continue her search for qualified women pilots. The initial stage of the Women's Auxiliary Ferrying Squadron (WAFS) was a *fait accompli.*

Technically, the WAFS did not exist as an activated unit. That was the nickname given to the women ferry pilots hired by the Ferrying Division in the fall of 1942 and stationed at Wilmington with the 2nd Ferrying Group. The name, however, stuck. Nancy Love's job was to supervise Squadron No. 1—the first squadron of twenty-five to fifty women. As other squadrons were built at other ferrying bases, other squadron leaders would be appointed.[22]

No one officer was responsible for male pilots employed by ATC. The male pilots reported to the commander of their Ferrying Group (FG) and the FG Commanders reported to the Commanding Officer. The women were to be organized the same way. No one woman would be directly in charge of all women pilots. The women of Squadron #1 would report to Nancy Love, who reported to 2nd FG Commander Colonel Baker, who reported to Colonel Tunner.[23]

Things were changing. Having women in the U.S. armed services was new. Now women would be flying Army airplanes. What next?

***

Jacqueline Cochran arrived home from England on September 10, 1942. The following morning, she was greeted by the headline in the New

York papers that Nancy Love had been appointed to head the Army's new women ferry pilots organization. Cochran's subsequent handling of none other than Army Air Forces General Henry H. "Hap" Arnold is nothing short of remarkable. Hap was no pushover and was accustomed to getting his way. So was Cochran. Simply put, on September 11, she flew to Washington, confronted Hap, stamped her foot, and called in what, in her mind, was her IOU.

Cochran claimed that Arnold had promised her the command of a women's pilot group. She claimed that while Arnold was in England during the summer of 1942, he had told her, "I want you to come home to form a group of women pilots." [24]

Arnold only mentions Cochran's name once in his wartime memoir, *Global Mission.* He was in England May 1942 to meet with Prime Minister Winston Churchill and the top British military command. "Among other people with whom I talked was Miss Jackie Cochran, who was in Britain trying to establish a women's ferry command for the R.A.F. It was then we first talked over the idea of creating an organization of women pilots in the United States Air Force."[25]

Arnold's explanation to Cochran was that he had "given instructions to prepare and submit plans for use of women pilots but that he had not seen the plan that had apparently been activated and that the announcement had been made without his knowledge and was not in accordance with his own intentions."[26]

"Cochran felt wronged," historian Deborah G. Douglas, writes. "She met with Arnold ... and it was not an exercise in diplomacy; it was a display of rage and raw political power. It was clear to Arnold that Cochran was not playing games and that he had to make good on his promise."[27]

"The use of a few of our women pilots to ferry trainer planes is just one segment of a larger job to be done," Cochran states in her September 11 memo to Arnold. Failing to properly coordinate all the women pilot resources would be wasteful, she claimed. And she reminded Arnold,

"The top job is what you told me I would do and is the one I have been preparing to do during the past year. ... The announcement made yesterday will throw this larger plan into confusion unless you clarify immediately."[28]

Handwritten below the typed contents is the following: "It was this broader phase President Roosevelt had in mind when he sent me to you last summer [1941] on his own initiative."[29]

Hap Arnold knew he was caught. Here was a commanding general, overwhelmed with the gravity of fighting a war that—in September 1942—the United States and its Allies were losing. He was trying mightily to deploy his air resources in the optimum manner. In an instant, he made his decision.

He called General George and George's chief of staff, Col. C.R. Smith, to his office and, in Cochran's presence, told them that the women's pilot project should be revised and that they should work it out with her because she knew his views.[30]

Cochran maintained the FD/ATC had "slipped" one over on General Arnold with the announcement about Nancy Love and the formation of a women's ferrying squadron. She claimed the public announcement was made without his knowledge. The record, however, says that General Arnold acted on General George's memorandum of September 5, which fully explained the plan.[31]

General George's response—dated the following day, September 12—suggested that the ATC's plan to employ approximately fifty experienced women ferry pilots be retained. Training of additional women pilots would be the job of the Army's Flight Training Command. The ATC "did not have the facilities," and the Training Command did. The basic flight-training program used for male cadets "is readily adaptable to the requirement of the women's flight training program. After graduation, women pilots should be assigned to or employed by ATC for flight duty. After assignment to ATC, they would have no further tie with or responsibility to the training program."[32] Jacqueline Cochran would direct the women's pilot training program.

Photo 4. Jacqueline Cochran, Director of Women Pilots.

Photo courtesy of Coachella Valley History Museum

Arnold sent a memo to General George E. Stratemeyer, Chief of the Air Staff, dated September 13, suggesting that Jacqueline Cochran be designated Director of Women's Flying Training. Arnold asked that a plan be presented to him "not later than Wednesday."[33] Colonel Luke Smith, Training Command, sent a proposed plan, "Women's Flying Training," to General Arnold, dated September 15, 1942.[34]

On September 15, Arnold announced that Cochran would form and lead a second women's program, the Women's Flying Training Detachment (WFTD), to train more women pilots to ferry airplanes for the FD/ATC. That same day, on General Arnold's instructions, Cochran met with representatives of the ATC, the Air Staff, and the Civil Aeronautics Association (CAA) to develop a plan based on General George's memo.

Colonel Tunner did not realize a big storm was brewing and sent a lowly captain as the ATC's representative to the September 15 meeting. Capt. James I. Teague was outranked at every turn. Every other man at the meeting held the rank of colonel or higher. But he was a very astute junior officer. As the group, with Cochran's dogged input, hammered out the plan for the training and use of women pilots, Teague became increasingly uneasy about the proceedings. He reported his concerns to Colonel Tunner.

Colonel Tunner, in turn, reported to General George. He summarized the suggested plan for use of the WFTD pilots in writing, and added a critical sentence: "[Graduates] will be employed at Headquarters, 2nd Ferrying Group, Ferrying Division, ATC ... only if they meet the basic requirements for the position of women civilian pilots (including physical examination and flight test) and not because they have graduated from the course outlined above."[35] Tunner believed he had the right to turn down WFTD graduates if he felt they didn't meet Ferrying Division standards. General George, fearing General Arnold's wrath, continually cautioned Tunner to go easy in that direction.

Another bone of contention was a "director of women pilots." Nancy Love had been appointed director of a women's squadron. The Ferrying

Division had no need for a "director of women pilots" as the women added to squadrons later would be serving under the Ferrying Division sector commander where they were assigned. General George's memo to General Arnold clearly stated that. The women pilots would be organized into squadrons of fifty. Nancy Love would "supervise" the first squadron and she would report directly to the Eastern Sector Commander. Other women would be appointed to lead subsequent squadrons as they were activated and those leaders would report to their appropriate Sector Commander.[36] Additional squadrons would be formed elsewhere across the country, but it would take until early 1943 to make them operational.

This was not to Cochran's liking and, by now, she had it in for the ATC and the Ferrying Division. She wanted to be director of *all* the women pilots flying for the Army and have them organized into a separate women's corps—answerable to her. She began to look for ways to accomplish that.

Teague observed the back and forth of communications in the days following the September 15 meeting, then drafted a memo to Tunner, dated September 22, expressing his growing concerns: "I have a definite feeling of suspicion in regard to the actions of Miss Jacqueline Cochran. This is principally a hunch, but I am convinced that we should probably do something about it, and convey to General Arnold our attitude toward her." Teague describes how Cochran, by innuendo, made it plain in the meeting that she "considered herself the only person who could efficiently be in charge of the Women Ferry Pilots."[37]

Teague also reports that he was asked to agree to have "one of our flight officers or transition officers at Wilmington give the flight check, and not to let Mrs. Love give a flight check on a woman who had already been flight checked by Miss Cochran. ... Miss Cochran, as far as the public is concerned, is coming to us and bringing us women who have been trained, and we should be appreciative. I, on the other hand, fear the Greeks bearing gifts."

Yet another contention Teague encountered was the number of hours recruits to the training program would need. The Air Transport Command wanted a minimum of 300 hours' flying time to qualify for acceptance into the WFTD program, the same number required for male pilots. ATC lost that battle. The compromise was 200, though Cochran said at the time she didn't think there were more than 150 women with 200 hours and that the ATC's demands were "ridiculous." She later denied using the word ridiculous.[38]

Arnold had a war to win. He took the easy way out, let the ATC take the blame (in Cochran's eyes) for what had happened even though he had full knowledge of George and Tunner's efforts throughout the summer of 1942 to form a women's ferrying squadron.

Arnold had surrounded himself with bright young men. His instructions to them: "Think of the problems confronting us. Think of the solutions to those problems. Bring in new ideas. Don't get mixed up with the routine operations of this office. Think of the future of the Air Force! ... I turned people loose and forgot about them." The Air Transport Command was a case in point. "Once it had established its bases around the world, and General George and General C.R. Smith were operating it, I was able to forget about it...."[39]

Hap Arnold knew exactly what he was doing.

Cochran accepted Arnold's explanation, took it very much to heart, and never forgot it, never let it rest. From then on, she blamed the ATC, General George, C.R. Smith, Colonel Tunner, Nancy Love, and her husband Colonel Love for what she perceived as an attempt to wrest from her the leadership of the women pilots that she felt was hers by right.[40]

Aviation writer Rowland Carter wrote, in depth, about Love and the formation of the WAFS and Jacqueline Cochran's Army-sponsored women's flight school in the December 1942 issue of *Flying Magazine.*[41] "In March of this year, with greater official encouragement, Mrs. Love quietly began laying the groundwork for the WAFS by beginning a study of ferrying military aircraft, within the operations section of the former Ferrying Command. ... Arrangements had been worked out in detail when Secretary Stimson came out with his announcement."

General George pointed out in the *Flying Magazine* article, "If they [the women pilots] show they can fly four-engine bombers safely after proper periods of training and preliminary work, I see no reason right now why they may not get the chance." A War Department spokesperson was quoted as saying that the women pilots who would undergo flight training at the Army facility Cochran was working to set up "will be trained in cross-country flying, using all navigational aids, to qualify as operating personnel for Mrs. Love's and such other units as may require their skills."

On October 2, 1942, Nancy Love received a letter from her boss from her Air Marking Days, John S. Wynne. He wrote:[42]

> I think you have a very important, interesting assignment and responsibility. Unquestionably you have had excellent training and background that convinces me you will make good in a big way. ...
>
> I used to ask you what you expected to do with your flying. ... So many times I have gone against the advice of my friends and have been gratified that my judgment was better than theirs. In your case I think you ought to be congratulated on your singleness of purpose in disregarding the advice of many of your good friends including me who used to try to minimize your deep interest in active flying. I stand corrected. ... I am very much interested in what you are doing and its success.
>
> Sincerely, your old boss, Jack

Chapter 3

Recruiting the WAFS

Building a squadron of experienced women pilots to ferry airplanes for the Army was not an easy task. The eighty-three recruitment telegrams Nancy Love sent out on September 5, 1942, were to women she believed to be qualified, but the women were slow to respond. Several who were interested were under contract teaching future Army and Navy combat pilots how to fly and were not able to get away immediately.

On Monday, September 14, eight days after the telegrams had been delivered, five women had been accepted. Betty Gillies was the first to report and qualify; Cornelia Fort was the second, followed by Aline "Pat" Rhonie, Helen Mary Clark, and Catherine Slocum. Adela Scharr, Esther Nelson, and Teresa James followed them.

The women would be part of the 2nd Ferrying Group, New Castle Army Air Base (AAB) in Wilmington, Delaware. In preparation, Colonel Baker put out a memorandum on September 21: The women would be designated as Civilian Pilots. They would have all the privileges of officers including use of the Officers' Club and the Officers' Mess. They would stand formation and roll call along with the men at 8 a.m. each morning. And they would live in BOQ 14 during their indoctrination period.[1]

BOQ 14 was ready for occupancy September 18. It was a two-story wooden barracks containing 44 square-shaped rooms. The women were issued an iron cot, Army blankets, maple bureaus, a large pine wardrobe, and blue scatter rugs. Walls were knocked out on each floor to provide a lounge. In addition, BOQ 14 soon had the only Venetian blinds on the base—providing welcome privacy for the female occupants who did not like undressing in the dark. The bathrooms had open shower stalls and commodes, but were altered to offer some privacy. Washbasins lined one wall. A pine shelf above, with mirrors, offered a place to put toiletry items. This would be home to the original WAFS and to all the women assigned to ferry airplanes for the 2nd Ferrying Group.[2]

Though the WAFS were experienced pilots, they knew nothing about the military or Army flying. They received thirty days of orientation during which they checked out on the Army trainers and attended classes in navigation, meteorology, military courtesy, military law, and Morse code. They learned about weapons and guard duty and were given instrument training in the Link Trainer.[3] Once they passed orientation, they were assigned to ferry liaison airplanes.[4]

The WAFS' first ferrying assignment took place October 22–23, 1942. Nancy Love appointed Betty Gillies to lead a flight of six L4-B Cubs from the Piper factory in Lock Haven, Pennsylvania, to Mitchel Field on Long Island. (The L-4B was a 65-horsepower, single-engine liaison plane.) Fort, Clark, Rhonie, Scharr, and James made up her team.

On November 1, when the WAFS squadron stood at twenty, ferrying began in earnest. Eight to ten women at a time would go out on a trip. They flew in groups because they were ferrying slower airplanes that needed to refuel frequently and several were destined for the same base.

With the arrival of November came a change for the worse in weather. If the weather was particularly bad, it took them several days to get the planes to their destination. Each night they were out ferrying, the women had to RON (Remain Over Night) at a military facility en route to their final destination. They reported their location and status back to Operations in Wilmington.

During her first ferrying trip, November 11–14, Helen Richards—the tenth woman pilot accepted—was one of seven WAFS pilots assigned to ferry a Cub from the Piper plant in Lock Haven, Pennsylvania, to bases in the South. They picked up the airplanes as a group, flew partway together, then scattered to deliver them to individual destinations. The weather caused problems the first day. The field at Charlottesville, Virginia, had been underwater for three weeks and was a sea of mud. When the seven WAFS landed there, the airplanes flown by Barbara Poole and Fort nosed over, damaging their props. Richards managed to land safely, after which she taxied to the parking area, cut the switch, cleaned up the aircraft and climbed out—only to sink up to her knees in mud.[5] The next day, Richards delivered to Selma, Alabama, while the other four with flyable planes proceeded to Jackson, Mississippi, and New Orleans.[6]

Though they began by ferrying Cubs, the WAFS' primary assignment was to ferry the open-cockpit PT-19 Fairchild single-engine, 175-horsepower trainers built at nearby Hagerstown, Maryland, to Army flight training bases in the South.

On November 22, eleven WAFS pilots were dispatched on their first PT-19 delivery. Betty Gillies and Del Scharr were designated flight leaders. En route by train to the Fairchild factory in Hagerstown, they missed the bus in Baltimore by two minutes and, since there wasn't another until seven the next morning, slept on benches in the station, propped against their parachutes and B-4 bags. Their tardy arrival in Hagerstown brought the wrath of the Army down on them.

"Where the hell have you been?" demanded the captain in charge. "Everyone in the Army has been looking for you," Del Scharr writes. "No one cowed Betty and neither did she kowtow to anyone." All of five feet, one-and-a-half inches, Betty looked the captain right in the eye and calmly stated what happened—seemingly to deaf ears. To placate the unreasonable captain, she wrote a letter of explanation and sent it to Colonel Baker back at Wilmington. Already the weather had closed in, so they couldn't have left anyway. They spent that night and the next in a Hagerstown hotel.[7]

Although the PT-19 boasted greater fuel capacity and range than the Cub, its open cockpit was a detriment. By late November, that made for frigid flying, even with winter flight gear. The WAFS took off from Hagerstown on November 25, the day before Thanksgiving, in soupy weather, finally delivering to Union City, Tennessee, on November 27. By November 28 they were back in Wilmington. For the women of Nancy Love's ferrying squadron, the nomadic life of a ferry pilot had only just begun.

"I was so tired after that trip, I thought I was getting sick," Nancy Batson recalled. "I wanted to curl up in my bunk and die. And the next morning, they told us we had to go right back out again. That was when I realized I had a problem. I might not be able to take this kind of schedule. But somewhere deep inside me I knew I had to do it and I knew I could do it. I'd just have to find a way."

With that, Nancy discovered a secret—the secret that would get her through twenty-six months of ferrying airplanes under all kinds of conditions. "I learned to fall asleep at a moment's notice. I just lay down on a bench or on my parachute and B-4 bag and conked out. That saved me."

Nancy Batson's ability to sleep anywhere under any conditions became legendary in the Ferrying Division.[8]

Nancy Love achieved her target number of twenty-five when Dorothy Scott arrived on November 21. Twenty-two-year-old Dorothy earned her 200-horsepower rating just before she left for Wilmington, and also logged the final hours she needed in her logbook. She had 504.

In December 1942, the Ferrying Division began making plans for incorporating those new ferry-pilots-in-training in Texas into the WAFS. Colonel Tunner sent Nancy Love and several Ferrying Division officers to visit the other Ferrying bases to determine if they could—would—accept a women's squadron. If the answer was yes, what kind of arrangements were needed to house and feed those women? The contingent left Washington on December 8. Nancy—with Colonel Tunner's approval—was beginning her transition into any military airplane she felt she could handle. She flew copilot for part of the flight in the Lockheed C-36, twin-engine (450-horsepower each) cargo/transport.

That party arrived in Great Falls, Montana, on December 11, the day before a flight of six Stearman PT-17s, piloted by WAFS, was due to take off for Jackson, Tennessee. Nancy and the officers were there to talk to the command of the 7th Ferrying Group at Gore Field. While in Great Falls, Nancy met with her six WAFS pilots and briefed them on the changes that soon would take place. They were being split up. A few would go to each of the Ferrying Groups who opted to accept women ferry pilots. There, they would form the core of the new squadrons. The first graduates from the school located in Houston would arrive in the spring and augment the numbers of each of the squadrons.

Nancy and the male officers left Great Falls on December 13. They flew to Long Beach, home of the 6th Ferrying Group where Nancy checked out in the single-engine, 450-horsepower BT-13 basic trainer built at the nearby Vultee factory. On December 16, she flew an hour and forty-five minutes as pilot in command and two hours as copilot on the C-36 on the flight from Phoenix to Dallas, where she met with officials from the 5th Ferrying Group at Love Field.[9]

From Love Field, Nancy went on to Memphis to talk to the 4th Ferrying Group. On December 20, she was in Romulus, Michigan—Wayne County Airport—where she conferred with the command of the 3rd Ferrying Group. She also took a huge step forward in her aviation career. She flew for the first time as a copilot in a twin-engine B-25—two 1700-horsepower Wright engines, built by North American.

Back in Washington, Nancy and the Ferrying Division officials compiled what they had learned. As 1943 dawned, women's ferrying squadrons would be formed at Love Field in Dallas under the 5th Ferrying Group;

at Romulus under the 3rd Ferrying Group; and at Long Beach under the 6th Ferrying Group—in that order. Florene Miller would command the Dallas contingent with Helen Richards, Dorothy Scott, Betsy Ferguson, and Nancy, herself. The women would ferry Stearman PT-17s and other aircraft built in Wichita, Kansas, as well as North American AT-6s built in Dallas.

Nancy put Florene in charge in Dallas because she had bigger things in mind and would be stationed there only briefly. She planned to check out on every available aircraft in an effort to pave the way for her women to fly any and everything in the Army's aviation arsenal. She would be making trips to Romulus and Long Beach to help establish the squadrons there. After that, she would travel between units to check periodically on her squadron commanders and their charges. Now she truly was senior squadron leader.

Del Scharr would command the Romulus group consisting of Barbara Donahue, Barbara Poole, Katherine Rawls Thompson, and Phyllis Burchfield. Lenore McElroy joined the group in January. There, the WAFS would begin ferrying L-2Bs (Taylorcraft, single-engine 65-horsepower), L-5s (Vultee, single-engine, 185-horsepower), PT-23s (single-engine, 220-horsepower), and PT-26s (Fairchild PT-19s with enclosed cockpits).

Leader of the Long Beach contingent would be Barbara Jane (B.J.) Erickson, and stationed with her would be Cornelia Fort, Evelyn Sharp, Barbara Towne, and Bernice Batten. The women of Long Beach would ferry BT-13s and 15s built at the nearby Vultee factory. The remaining ten "Originals" would make up the Wilmington women's squadron with Betty Gillies in command. Helen Mary Clark would be her executive officer. They would continue to ferry the Fairchild PT-19s and PT-26s.

As 1942 ended, life for the WAFS was about to change dramatically.

While Nancy was out securing the future of the women of the Ferrying Division, the six WAFS she met in Great Falls—Teresa James, Nancy Batson, Delphine Bohn, Florene Miller, Phyllis Burchfield, and Kay Rawls

Thompson—were encountering challenge after unexpected challenge in their mission to take six open cockpit primary trainers from Montana to Tennessee in the dead of winter.

Six yellow bi-wing Stearman PT-17s—wings and tail surfaces encrusted with rime ice, open cockpits dusted with snow—stood as solitary sentinels in the pale December morning light. The thermometer on the side of the Operations Office, Great Falls AAB, read nine degrees Fahrenheit. Six figures, more closely resembling overstuffed penguins than human beings, stood silently on the tarmac. Six pairs of eyes gazed at the steeds they were scheduled to ride south. Snowflakes, carried on the northwest wind out of Alberta, stung their faces. The shivers that rolled across their skins, underneath the bulky flight gear, were induced by both wind chill and apprehension.

Teresa James was the flight leader for the six WAFS assigned to take those six airplanes from Great Falls, Montana, to Jackson, Tennessee. The date was December 12, 1942.

The nine degrees was compounded by the five-thousand-foot elevation at Great Falls. It was so cold the line crew had to heat the engine oil in order to get the Stearman engines started. They would be flying several thousand feet higher in that mountainous country with sharp, rocky ridges and few hospitable places to put an airplane down in an emergency. Primary trainers carried no radios. The pilots could not communicate with each other or call for help. Hand signals had to suffice.

In retrospect and during an interview in 2005, Phyllis Burchfield Fulton wondered aloud over the nineteen-day adventure those six endured when they were plagued by colds, flu, near pneumonia, several minor mishaps, plus the unpredictability of the weather in December.[10] "Sitting here sixty-three years later, on a bitter cold day in Ohio, I still remember the agony of the flight. Flying over the Rockies in an open-cockpit plane in December was sheer madness. Whoever planned that flight, sitting in a cozy office, should be thankful he didn't have to fly it. Only by a miracle did all six of us survive. I think we had six angels guarding us."

Army-issue bulky winter flying gear was a challenge. First, the women donned long scratchy woolen underwear and layers of socks. Over the long johns, they pulled on high-waisted, fleece-lined leather pants. These zipped from the shinbone of one leg up to the sternum and were held in place by suspenders. They topped that with fleece-lined leather jackets. Leather flying caps with chin straps and goggles, fleece-lined leather gloves, and wool-lined boots completed their standard flying garb.

"We thought, surely, all this would be enough to keep the cold out, but it wasn't," Alabama-born Nancy Batson remembered. "To that, we added our parachute, with straps over the shoulders and around the thighs."

For the flight out of Great Falls, the women also received chamois masks held in place by wide black elastic bands. The masks were to provide at least some semblance of protection against frostbite since they would be flying another few thousand feet above the earth where the air was even colder. The WAFS—used to flying open cockpit PT-19s in the cold, damp East Coast weather and at much lower altitudes—already had learned that when flying in open cockpit trainers, a runny nose was a constant companion.

Departing at the same time were twenty-seven male pilots who were ferrying the other twenty-seven PT-17s to Jackson.

Snow covered everything. The world was white. They couldn't make out highways or railroads or rivers. Even though the airplanes were yellow, they were hard to see. "You couldn't see the airplane ahead of you if you were level with it," Teresa wrote.[11]

The first day out, the wind tore the maps from the gloved hands of the navigator, Florene Miller. The six, lost and without radio communication, found a plowed-but-icy runway of an auxiliary field on which to land, confer, get their bearings and get back on course. They couldn't turn off their engines because they could never have restarted them.

Wyoming and Colorado were uneventful until they had to climb to ten thousand feet to get over eight-thousand-foot Raton Pass, the border between Colorado and New Mexico. "The motors kept cutting out and we couldn't imagine what caused it," Teresa wrote. "Believe me, if we'd been called on to make a forced landing, I'm not real sure where we could have put those ships down in that rugged terrain."

The trip became a comedy of errors. In New Mexico, they accidentally picked up some of the new 90-octane gas—they referred to it as "bad gas"—which caused their engines to stop and then restart even flying at the lower altitudes of West Texas. Then a mechanic stuck his foot through Burchfield's wing and that meant costly time out for repairs.

Photo 5. Six Original WAFS.

Signed photo from author Sarah Byrn Rickman's personal collection
Katherine Rawls Thompson, Phyllis Burchfield, Nancy Batson, Delphine Bohn, Florene Miller, and Teresa James and the six Stearman PT-17s they ferried from Great Falls, MT, to Jackson, TN, in December 1942.

Away from the frigid north country, instead they encountered unstable, Gulf-induced air over Texas and across the South. That meant rain, fog, and low ceilings and lower visibility that kept them grounded for days at a time.

"Everywhere we stop, we sit for at least three days," Teresa wrote.

They made Little Rock by Christmas Eve, got weathered in again and spent Christmas Day there in balmy but foggy 80-degree weather. They made it to Memphis on the twenty-sixth, only to get weathered in for several more days. Rain! Throughout the trip, bouts with head colds, flu, and tonsillitis plagued the group.

Finally, December 31, 1942, they delivered their aircraft to Jackson, Tennessee.

They didn't get away from Memphis until 1:10 p.m. because of a low ceiling, but they delivered the ships to Jackson, Tennessee, where they were met with, "Well, did you finally get here?"

"Turns out we beat all the men in," Teresa wrote. "Out of twenty-seven guys and six gals, we are the first ones to arrive out of that whole contingent that left Great Falls what seems like ten years ago. How 'bout these gals!"

They caught the American Airlines flight out of Memphis that night —New Year's Eve.

"One of the loveliest scenes flying is skimming over the clouds with the moon shining on them. It's like looking at silver balls of cotton. It's an awesome sight regardless of how many times you see it. Then coming into Washington D.C., I watched the sun come up. It's one of those moments that make you feel it's great to be alive!"

The six hopped a train for Philadelphia early in the morning New Year's Day, and a cab from there to the base at Wilmington. They had been gone for twenty-six days, most of the month of December.

***

The women's squadron remained at twenty-five until the last day of 1942. Pat Rhonie was discharged December 31. She had spent an additional night away from base after Colonel Baker told her to return. The WAFS were under a ninety-day, renewable contract that was expiring the end of the year. In an interview after the war, Nancy Love verified that Rhonie was discharged. "This was done through normal civil service procedures. She was given a hearing before a board first. Male civilian pilots were discharged in this way."[12]

Aline Rhonie was thirty-three and a well-known artist/aviatrix when she became the fourth woman to join Nancy Love's squadron. She had learned to fly in 1930, and by the time she joined the WAFS she had accrued 2,627 hours. After leaving the WAFS, she went to England and flew with the Air Transport Auxiliary (ATA).[13]

Joining on January 3, 1943, would be Helen "Little Mac" McGilvery and Kathryn "Sis" Bernheim. On January 23, 1943, Lenore "Mac" McElroy became the final member of the original WAFS. On January 26, General Arnold ordered that, from that date on, women admitted to Nancy Love's Ferrying Squadron must be graduates of the Army's flight training school in Texas. Though twenty-eight are credited as original WAFS, the squadron settled in at twenty-seven in late January and remained there until other circumstances changed its makeup.

The twenty-eight WAFS in the order they joined were: Nancy Love, Betty Gillies, Cornelia Fort, Aline "Pat" Rhonie, Helen Mary Clark, Adela "Del" Scharr, Esther Nelson (Carpenter), Teresa James (Martin), Barbara Poole (Shoemaker), Helen Richards (Prosser), Barbara Towne (Dickson Fasken), Gertrude Meserve (Tubbs LeValley), Florene Miller (Watson), Barbara Jane "B.J." Erickson (London), Delphine Bohn, Barbara Donahue (Ross), Evelyn Sharp, Phyllis Burchfield (Fulton), Esther Manning Rathsfelder (Shively, Westervelt), Nancy Batson (Crews), Katherine "Kay" Rawls Thompson, Dorothy Fulton (Slinn), Opal "Betsy" Ferguson (Woodward), Bernice Batten, Dorothy Scott, Helen Schmidt McGilvery, Sis

Bernheim (Fine), and Lenore McElroy. Catherine Slocum qualified but immediately resigned. She had young children and was needed at home. She never ferried; therefore she is not counted in the original squadron.

The women were issued khaki flight coveralls (made for men, therefore far too large), a parachute, goggles, a white silk AAF flying scarf and leather flying jackets with the ATC patch. Eventually they would acquire men's long woolen underwear.

Nancy Love chose the WAFS uniform. The jacket was gray gabardine, belted, with four patch pockets, squared shoulders, straight set-in sleeves and brass buttons. With it, the women could wear a gored skirt or slim-cut slacks of the same material. Shirts and ties were of contrasting tan broadcloth. Brown leather low-heeled pumps, a brown leather shoulder bag, an overseas cap, overcoat with removable lining, and gloves completed the outfit. To that they added the command insignia—the civilian pilot wings of the Air Transport Command—over the left breast pocket.[14] The women paid for their own uniforms.

Once they had uniforms and a sufficient number of girls to practice close-order drill, Colonel Baker ordered the WAFS to march in review every Saturday morning with the rest of the troops stationed on base. Delphine Bohn relates that Nancy Love was not adept at leading others in marching formation. "Her voice was too soft and her commands too tentative and the women frequently got orders mixed up and went off in the wrong direction laughing uproariously. However, they all survived."[15]

Nancy did keep a close eye on her girls' behavior. She was the squadron's disciplinarian and initially had to act as housemother. It wasn't long before she and Colonel Baker hired Mrs. Margaret Anderson (Andy) to fulfill the latter function.

Once the press got wind of the fact that women were being hired to fly airplanes for the Army, New Castle AAB was deluged with reporters and photographers and requests for interviews. This, it seemed, was the story of the century. Nancy Love hated the limelight and sought, if anything, to avoid publicity. Likewise, the ATC tried to shield her and her WAFS, but this was a big story and the press played it to the hilt.

Many of the stories written and disseminated were blatantly incorrect. Some were out-and-out fabrications. The women were under not only the bright glare of publicity, but the microscope of public opinion as well.[16]

Following several requests from magazines for ghostwritten accounts of WAFS on duty, Colonel Tunner put a ban on that type of publicity. "Stories of this type are not considered to be for the best interests of the Ferrying Division and will tend to over-glamorize the members of the WAFS."[17] But the word was already out.

Women joining any of the military or military-auxiliary services found themselves under a microscope. Negative comments flourished. Women disapproved, men disapproved. Both were quite vocal. Nevertheless, women were part of the effort to fight the war.

Air Force Major General Jeanne Holm writes of the prevailing attitudes in 1942 when the Women's Army Corps (as well as the WAFS) was being considered: "This notion that women should be of high moral character and technical competence while no such standards were used for men set the tone for the double standards that were to characterize the women's [military] programs for the next forty years."[18]

Part II

Building the WFTD

October 1942–April 1943

By October 1, 1942, efforts were underway to teach women on the Home Front trades like welding, armature winding, and burning to make them eligible for work in America's shipyards. Republic Steel had hired 1,000 women to make and assemble aircraft parts and accessories. Grumman Aircraft began training women to be engineering aides. And "women were joining up in record numbers," the WAACs and WAVES reported.[1] Spokespersons from First Lady Eleanor Roosevelt on down were advocating women in the workforce as part of the country's war effort.

By late October, the WAFS numbered twenty. They made their first ferrying trip October 22-23, 1942, when six women took six L-4B, 65-horsepower, single-engine liaison aircraft—Piper Cubs—from Lock Haven, Pennsylvania, where they were built, to Mitchel Field on Long Island.

On November 16, 1942, the first twenty-nine women recruited by Jackie Cochran for the Women's Flying Training Detachment entered Army flying training at Houston (Texas) Municipal Airport. That class, known as 43-1, would graduate April 24, 1943, and be sent to one of the four now-active women's ferrying squadrons.

Women now were flying Army aircraft.

Chapter 4

The Women's Flying Training Detachment

While Nancy Love was building her squadron in Wilmington in September and October 1942, Jacqueline Cochran moved her headquarters from Washington to Fort Worth, Texas, to work with the Army Flying Training Command (FTC) to organize and implement the Women's Flying Training Detachment (WFTD).

Cochran's objective of forming a women's flying corps in the United States had gained impetus when she visited England in 1941 and again in 1942.

When the first civilian ferry pilots' pool was set up in England on January 1, 1940, eight experienced women flyers were included. One of them, Pauline Gower, already had approached the Director General of Civil Aviation at the Air Ministry, Sir Francis Shelmerdine, about using women pilots in the defense of England. She was chosen to lead the women of the Air Transport Auxiliary (ATA).[1] In June 1941, Cochran met with Gower and took the idea of women ferry pilots back to General Arnold, but Arnold told her this was not the right time to form such a group. He didn't think women could handle the Army's airplanes, even the small single-engine aircraft. He had plenty of men, he said, and no need to involve women in flying duties. But later that fall, Arnold suggested that Cochran recruit some of the American women pilots to fly with the ATA. Spring of 1942, she put twenty-three of the United States' best women pilots on a ship to England. Cochran herself flew over once they were set up and remained in England for several months while the women settled themselves.[2]

Her return to the U.S. on September 10, 1942, coincided with the announcement that the WAFS was being formed. Within days, Jacqueline Cochran was designated Director of the Women's Flying Training Program in the USAAF, with the status of a one-dollar-a-year employee —which meant she was compensated by a token salary of one dollar a year.[3] She would be in charge of selecting the women trainees.

A month later, October 7, General Arnold finalized the official authorization for training women pilots. Training was to begin "as early as practicable." The objective: the qualification of 500 women pilots for ferrying duty for the ATC not later than December 30, 1943. Training would begin on primary trainers with subsequent transition into basic and advanced trainers. One significant change: the candidates needed only a minimum of 75 hours' certified time to be accepted—not the 200 agreed to at the September meeting. Changes in hours needed would not be published and individuals would be selected based on their own qualifications.[4]

November 3, 1942, General Arnold went a step further. In a signed, updated directive to the Training Command, he stated that fighting the war was going to "tax the nation's manpower." He ordered "a maximum effort to train women pilots ... to place women in every noncombatant flying duty in which it is feasible"—even to the point of accepting for training women with *no* previous flying experience. This, in essence, eliminated—if Cochran saw fit—any stipulation of number of hours flown.[5]

This was a huge change. Men who had no previous flying experience were taken as cadets because if they washed out they could be sent to the walking army. This was not true of women. The women weren't military. If they washed out, they would be sent home. Nevertheless, General Arnold was now firmly in the women's camp, and he remained so.

On November 3, 1942, Jackie Cochran and Gen. Barton K. Yount—head of the Army Air Forces Flying Training Command in Fort Worth—selected Aviation Enterprises at Howard Hughes Field, Houston Municipal Airport (now William P. Hobby Airport), to be the home of the Women's Flying Training Detachment. They made that decision despite the lack of bathrooms and a dining hall in the Aviation Enterprises hangar. The women would have to march half a mile to the airport terminal for their meals and to use the bathroom facilities. There were no barracks. Living quarters for the girls were non-existent.[6]

Cochran badly needed someone she could depend on to ride herd on her twenty-nine incoming women pilots who made up Class 43-1. When she interviewed Dedie Deaton, she knew she had struck gold.

Leni Leoti Clark was born September 2, 1903, in Hereford, Texas, and moved with her family to Wichita Falls in 1915. She married Clifford Deaton on July 22, 1922, and they had a son, Clifford Jr.

When WWII began, the lady by then known as Dedie was comfortable. She was the wife of an insurance executive, active as a Red Cross field volunteer and swimming instructor, a budding executive, and a mother whose only child would soon go off to college. But World War II changed a lot of lives—Dedie's among them. Like original WAFS Betty Gillies and Helen Mary Clark, she went off to serve while her husband—declared too old for the Marine Corps in which he was to have a commission—stayed home.

When Deaton met Cochran in Fort Worth on November 8 for the interview, she had no idea that, in accepting the job, she would be away from home for more than two years. Cochran quickly arranged for a

Civil Service appointment for Deaton and named her staff executive. Dedie would be moving to Houston immediately, where her first task was to find housing for twenty-nine women, not an easy job in wartime Houston. Housing was at a premium. She also would be looking for transportation for the women to and from Howard Hughes Field.[7]

I went with confidence that whatever I did and the decision I made, she [Ms. Cochran] *would back me up, I just felt it. She said it. "You know more about this than I ever will. This is your school. You handle it your way and I'll support you." She said those words, and she did. The only thing that she didn't support me in—[Laughter]—as long as I did not get recognition and as long as I got no publicity, we got along. But let something happen to bring me into the limelight and she took it up. You see she wanted everybody to think that she ran that school and nobody else had anything to do with it. She wanted to have sole credit for anything good that happened. And I was the whipping boy to take the blame for anything that didn't go right. Hap Arnold was guilty of that too. They said I had to run the school; I had to make the decisions. Fortunately, the Lord looked after me.*[8]

When she arrived in Houston, Dedie discovered that she had an unexpected ally already in place. Lieutenant Alfred Fleishman, who had been serving in the Missouri State Guard, had received his commission in the USAAF and by fall 1942, the thirty-six-year-old St. Louis native was in command of the Ellington Army Air Base supply depot adjacent to the WFTD trainees' area. He knew about Dedie's impending arrival and why she was coming. Her first morning on the job, he called and offered her a ride out to what would be the WFTD facilities at Hughes. Using his own landlady as a resource, he also helped Dedie find her own living quarters.

Photo 6. Dedie Deaton, Chief Establishment Officer

Courtesy WASP Archive, Texas Woman's University, Denton, TX.

Fleishman made Dedie Deaton's transition to Houston and the 319th AAFWFTD (Army Air Forces Women's Flying Training Detachment) possible. He told her to call on him if he could do something else for her, and he meant it. Thus was forged the team that would make the WFTD succeed.

Photo 7. Lt. Alfred Fleishman

Courtesy WASP Archive, Texas Woman's University, Denton, TX.
Lt. Alfred Fleishman was the Physical Education Director for the Women's Flying Training Detachment (WFTD), Tactical Officer, and Special Services Officer, Houston, TX.

Attacking the housing dilemma, Deaton was able to get rooms at several tourist courts approximately ten miles from the airport, but they would not be available until December. She secured them for Class 43-2—due to report mid-December—then went about finding alternative lodging for the first class.

I had to take responsibility for the tourist courts. I got them! Jackie like to have fired me. She was just horrified! Motels had bad reputations then.[9]

There wasn't any precedent. I had no guidelines. Jackie Cochran said, "Go down there and take care of the girls." And I said, "By what authority?" She said, "the military has absolutely no authority as far as they're concerned. This is yours, you take care of them. Place them, get them a place to eat, see that they go to a beauty parlor once a week. See that they stay out at the airport and have study hall every night from 7:30 to 9:30."

Those poor things coming out there before breakfast—well, of course, they didn't have any place to eat at first. Then the contractors set up a mess hall. We had to be out there at seven o'clock. No breakfast, no coffee, no nothing.

They take their drill, do their PT, fly, and go to ground school in the same old sweaty coveralls. No place to shower. Two potties. [There was a second one in the Civil Air Patrol hangar.] *No place to sit down, nothing. And then stay in the mess hall for two hours that night to study. They did it for three nights and I never felt as sorry for anybody in my life. So I cut out the study hall. And when Cochran found out, she liked to have fired me. But that's just inhuman.*

I was supposed to police all of them, scattered all over town, and see that they didn't get out of their quarters after eight o'clock at night. And that they went to the beauty parlor every week.[10]

When W-1 was there, right after we got our infirmary, we had five gals that got pneumonia. They got it sitting on the wet ground and being out in the fog and stuff. Before the hospital would take them, I had to pay their [one week's] *hospital fee in cash out of my own pocket. Of course, Ms. Cochran reimbursed me later. And we couldn't have done that if it hadn't been for Ms. Dire our nurse. They called her the duchess. We all loved the duchess. She was something else!*[11]

The women were issued khaki coveralls in men's sizes 40 to 46. The trainees called them "zoot-suits" and learned to roll up sleeves and cuffs, cinch in belts, and to deal with a crotch that hung to their knees. The only requirement—they had to wear a hairnet on the flight line. No shoes or boots were issued. They wore what they had. To make matters worse, Captain Paul Garrett—the commanding officer of the 319th—took Dedie aside and told her, "Keep the girls happy and out of my hair." This same commanding officer later told the women trainees that they were civilians and were not obligated to follow Deaton's, Fleishman's or the WFTD's rules.[12]

Thirty-nine-year-old Mrs. Deaton was who she was—a small-town Texas housewife and Red Cross volunteer. Educated, yes; smart, yes; capable, yes; but dealing with the individualists of Class 43-1, the group she encountered at Houston's Rice Hotel on November 15, would stretch her patience and try her staying power. And she would be answerable to Jacqueline Cochran. However, Deaton was blessed with Texas grit and determination.

She backed me up, but she wasn't gonna let me be the one that they were to be grateful to. She was very careful to keep me in my menial place. One thing, I didn't give a damn if I got sent home.[13]

Deaton would come to terms with her charges and her job and she would exceed her own and Jackie Cochran's expectations. But on November 15, 1942, Dedie Deaton must have faced the immediacy of the present and all its demands with trepidation. Dedie would oversee eighteen classes of women ages eighteen-and-a-half to thirty-five, beginning with the 43-1 "Guinea Pigs" in Houston and ending in Sweetwater with the "Lost Last Class" of 44-10. That last class graduated December 7, 1944, thirteen days before the December 20 WASP dissolution date. She would see 1,074 women earn their wings and graduate. She would live through the deaths of eleven of her trainees. She would remain lifelong friends with many of her charges.

Chapter 5

Class 43-1 — The "Guinea Pigs"

The first class was nicknamed the "Guinea Pigs"—and for good reasons. The program was experimental and had several strikes against it—not the least of which was the Army was trying to kill the experiment before it could get off the ground. Nevertheless, the Guinea Pigs became legends in their own time. The twenty-three who graduated were the most experienced women flyers to enter training. Only the original WAFS —who were not sent through the Army's flight training program—had more flight experience. The average flying hours of the twenty-nine who entered training November 16, 1942, was 315 and one, who was an instructor, had 762.[1] In fact, several were qualified instructors.

***

Twenty-nine-year-old Jane Straughan had been flying for five years when she received an invitation to attend a cocktail party in Jacqueline Cochran's suite at Washington's Mayflower Hotel, in the fall of 1942.[2] Jane was a member of the Washington D.C. Chapter of the Ninety-Nines. Jane and her fellow chapter members had gathered to hear what Cochran, now the president of their organization, had to say.

Miss Cochran told us about the twenty-three American girls she had taken to England the previous spring. We were all very interested and impressed with the idea that they did that, then after a few more drinks, she asked how many of us would be willing to fly for our country? Of course everybody by then would fly for their country.

I heard nothing more until a telegram arrived one Sunday morning saying, "Report to the Rice Hotel in Houston, Texas, such and such a date at your own expense. I thought it was a joke until Mary Lou Colbert called and asked if I was going.

I had just gone to work for Engineering Research Corporation, which made the Ercoupe aircraft.[3] *I was assistant sales manager and didn't think I should leave that job. My husband, Al, was a Service Pilot and had just received a commission. He thought he would be sent overseas, so he said, "if you want to, go on down to Houston and find out what it is all about." Then my boss, the sales manager at ERCO, said that due to the shortage of materials, we probably wouldn't be making Ercoupes much longer and that he, too, thought I should go. So he kind of talked me into going.*

Mary Lou Colbert (Neale) was the twenty-eight-year-old daughter of Rear Admiral Leo Otis and Florentine Odou Colbert. [4] She had learned to fly through the CPT Program at Georgetown University. Mary Lou knew Nancy Love. They were the same age, they came from similar backgrounds, and both had attended an elite women's college in New England: Mary Lou, Wellesley and Nancy, Vassar. Mary Lou also knew Jackie Cochran. She was invited to Cochran's cocktail party for the Washington-area Ninety-Nines in September 1942.

Photo 8. Mary Lou Colbert, Class 43-1, Women's Flying Training Detachment (WFTD). Jacqueline Cochran's first recruit.

Courtesy WASP Archive, Texas Woman's University, Denton, TX.

I knew Jackie pretty well. Everybody knows about her upbringing, it was very tough. She never played up to anybody. I admired that. You sensed a kind of loneliness in her, too. She had not been raised with a lot of love at all, anything but.

Claire Callaghan and I used to talk about what would be the best. When we found out that Jackie really was going to have a cadet training school, that sounded better than what Nancy had. The training wasn't there. And that's what Dad wanted. When Nancy was starting, she was very insecure as to how it would come about. So she wanted five hundred hours. When I told her I had four-hundred-fifty, she said, "Quick, get some more."

When Miss Cochran came back from her meeting with General Arnold [that day in September 1942], *I was in the lobby waiting for her. "Okay," she said, "you're the first one, come on up and sign in."*

Jane, Mary Lou, Claire, and three others—Edna (Edie) Collins, Marjorie Ketcham, and Dorothy Young—went to Houston from Washington. All six would be among the twenty-three who graduated on April 24, 1943—the first class of women to receive and complete flight training through the Army's Flying Training Command.

They were Jackie's girls. Hand-picked, the first chosen, Class 43-1, "the Guinea Pigs." The other seventeen who would graduate with them were: Eleanor Boysen, Gertrude C. (Brownie) Brown, Byrd Howell Granger, Marjorie Gray, Evelyn Greenblatt, Ruth Hellman, Ann R. Johnson, Vega Johnson, Marjorie Kumler, Geri Lamphere, Marion Mackey, Margaret McCormick, Elizabeth McKinley, Sidney Miller, Lovelle Richards, Betty Tackaberry, and Magda Tacke. Twenty-nine entered training and twenty-three graduated.

Jane was, in her words, quite in awe of the women who gathered at Houston's Rice Hotel on November 15.

Such a variety of personalities, I could tell these girls had probably gone out and bought any little airplane they wanted and they had lots of time. I had just a little over 200 hours.

Dedie handed them a list of hotels, motels, rooming houses, and private residences that would take girls. The women would, of necessity, be scattered all over Houston. To get to the airport, the twenty-nine found transportation based on their individual financial state and who their friends were.

A friend of the Colberts had told Mary Lou that when newly appointed WAAC commander Oveta Culp Hobby left for Washington, she had turned her home over as a boarding house for young girls in Houston. [5] Mary Lou and her traveling companions looked it up.

That's how four of us ended up in that lovely home. Mack (Elizabeth) McKinley and I roomed together as did Claire Callaghan and Jane Straughan. People saw all these girls around and asked what we were, and that's when someone said, "oh, they're just a girls' basketball team." Which was obviously insane, because look at my size! I needed a pillow just to sit high enough in the cockpit."

Nothing had been done for the women in advance of their arrival. Men were still in training at Houston Municipal and they took precedence over a bunch of women pilots. Because the average age of the first class was twenty-eight, the women were knowledgable about living on their own, making a living and what to do with their money. They also were smart enough to know they needed some help in a strange city.

Again, Mary Lou's family connections paid off.

My father's good friend from Houston, Jesse Jones, the Secretary of Commerce, came to our aid. We needed a mutual bank account of some kind. We were getting paid and nobody knew where to deposit it. So, I got a letter from Jesse with his contacts and he set it up with a bank. The second thing we did was set up our own insurance, which this banker suggested we do. I was so glad, because I got very sick at one point and so did one of the other girls. We had insurance to pay for the hospitalization.

Because the women were civilians under Civil Service, they had no military benefits. They also lacked the military savvy Mary Lou had learned from her father.

Your hours didn't indicate how much you know about the rules and regulations. That's why this course had to be done. I think that's the problem Nancy's group ran into. If you have a private airplane, you know your own regulations, but there's a lot of things that, commercially, you won't know. Details we needed to know for the ferrying program, things you won't know unless you had done a lot of cross-country flying.

You have to learn meteorology. You might not know much about aerodynamics, so you're not sure how far you can push the button with some things. You have to know more. You have to know all the rules and regulations. You have to pass tests to graduate up from just private pilot to commercial pilot. That's why we needed the classes.

***

Betty Guild Tackaberry (Blake) also met Nancy Love before being recruited by Cochran. Her friend from flying days in Hawaii, Cornelia Fort—one of Nancy's original WAFS—introduced them. Betty and her family, living outside Honolulu, were eyewitnesses to the attack on Pearl Harbor. The young ensign who had brought Betty home from a party the night before had stayed the night at Betty's father's insistence. That saved the young man's life. The morning of December 7, his bunkmate was killed when the ship's cabin in which he slept took a direct hit. Betty married the ensign and when he was transferred stateside, she went with him.[6]

September 1942, I was living in Philadelphia. I got on the train and went down to Delaware to meet Cornelia and Nancy Love. I was very impressed with Nancy. She was a lovely looking gal. She was probably thirty, had this premature white hair, but it was stunning on her, and she had a beautiful tan. She was soft-voiced and had a finishing school girl feeling about her. I liked her very much, and I was very disappointed when she said they couldn't accept me because I didn't have a 200 horsepower rating. So very sadly I went back to Philadelphia.

Then Cochran sent me a letter. Was I interested in being in her first class? Of course I said, "Yes." She invited me down to Bolling Field, Washington

D.C., and we took our physicals there. That was October. In November, I went to Houston, the Rice Hotel. I put on my best clothes—a maroon suit and high-heel maroon alligator pumps and an alligator purse and I probably had gloves. There were two other girls there that were very well dressed, one from San Francisco and one from Texas. All the rest had tennis shoes on and sweats and Levis and looked very different.

After Jackie gave us a big talk she asked us to all just decide who we wanted to room with, and she gave us lists. Some private homes accepted people as roomers. A lot of girls did that. And there was a complex of apartments, little brick two-story apartments or condos or whatever you want to call them. We could rent one of those. Then some of the girls stayed in a hotel. One of them was an heiress and she and her roommate, who was a DuPont, stayed in a fancy hotel—for the whole six months—but we couldn't afford that. Ruth Hellman, Sidney Miller, and I were the only three well-dressed ones there, so we gravitated together and decided we would room together and rent an apartment. We rented furniture by the month, and we rented an old jalopy car, so we drove ourselves out to the field everyday.

Geri Lamphere (Nyman) graduated from high school at barely sixteen; she spent a couple of years at Washington State University, attended business school, and got a job at a radio station. Boredom spurred her on to take flying lessons and she immediately took to it. Her salary wasn't sufficient to pay for lessons, but her instructor needed a ground school instructor. He told her that if she would get her ground ratings and teach ground school for him, he'd give her free flying lessons. A quick study, Geri was teaching in no time.[7]

Within a year I had all five ground ratings including A&E,[8] my private, my commercial and my instructor's ratings. I made good money teaching flying and went in business with the instructor. But Pearl Harbor put a stop to flying along the Pacific Coast. I was out of a job.

Photo 9. Geri Lamphere, Class 43-1.

Courtesy WASP Archive, Texas Woman's University, Denton, TX.

I wrote to the Army Air Corps. They kept saying "we need pilots, we need pilots." Here I was with a lot of hours. But I got so disgusted with them because I kept getting a form letter back saying, "We do not feel that women will be capable of flying military aircraft." Then someone told me about Jackie Cochran, that she had taken a group of women pilots to England and might be interested in taking a second group over.

I wrote to her and said, "They don't want me here, I'll be happy to go to England." She wrote back that with my hours and experience, I qualified. Then, two weeks before I was to go to England, I got a call from her—not her secretary, but directly from her. "I need you to come to my apartment, I need to see you." Miss Cochran was bed-ridden, hooked up to an oxygen tank. Spare tanks were in every room.

"I've just had authorization to form a women's flying group," she told me. "I need twenty-five with lots of hours, but mostly I need them with lots of stamina."

Then she showed me a letter. "I just had this from General Arnold," she said. "It is not a letter that I was supposed to get, and probably nobody would say that it was ever written." It was a letter from the War Department to the commanding officer at Houston where the first class was assigned. The main thing I remember about the letter was it said, "we know we must give these women an opportunity to fly. We do not think they will ever be able to fly military aircraft. Get rid of them as soon as possible."

Miss Cochran said, "That's why I need women with lots of stamina."

I don't know if some of the other girls were shown this letter or not. Or whether I was the first one she interviewed. I have no idea. But she said to me, "If this first class doesn't make it, the whole thing will go down the drain."

Geri got the names of three others going to Houston from New York. She, Byrd Howell Granger, Magda Tacke, and Marion Mackey drove two cars to Houston. Once there, Geri and Magda decided to room together. Marion and Dottie Young from Washington D.C. hooked up and lived near them. The four drove to the airport together every day.

The Houston airport was something else. The women were to occupy an old building, with no bathroom, at the far end of the field. They had to trudge a half mile each way on a hard-packed shell road to get to the commercial terminal and a ladies' room. The process tore up many a pair of shoes—which were rationed and therefore hard to replace. They did eventually get a portable outhouse out by their remote building.

Their headquarters and classroom building was an old shack at the end of the field. That's where they did their studying and training. In order to eat, they had to walk that mile to and back from the terminal. Geri recalls that they were getting starchy foods that were really bad. "We asked for salads for lunch. So they gave us beans on a lettuce leaf. To show their good humor, they gave us brown beans one day and white the next."[9] Finally food was brought in for them—though it wasn't very good.

The commanding officer I'm sure, having gotten that letter, was determined that these girls were not going to make it. But one instructor was good to us. We called him Pappy. He was an old-time instructor, not one of these new bright youngsters that had just graduated.

No military aircraft were waiting for them at Houston Municipal Airport. Already experienced civilian pilots, the women had to learn to fly "the Army way" in a hodgepodge of civilian aircraft. They flew Stinsons, Fairchilds, Cubs, Reliants, Aeroncas—whatever had been scrounged up for their use. As it turned out, the civilian airplanes were foreign to those newly minted instructors who had just been trained in military planes. But Geri and the other women had learned to fly on those civilian aircraft.

I think that's what saved us. We all had been flying those airplanes. Many of the girls had been instructors before joining and had more time and experience than some of the instructors—who were civilians, too. We were able to show them how to operate the planes.

One of our girls was a stunt pilot. When she got aboard for her qualifying run, her young instructor asked her if she could really fly. She answered, "Do you really want to know?" She took off down the field, turned and flew back over the field upside down, righted and flew straight up in the air until she stalled out. What a performance! The young lieutenant staggered off the plane and vowed he would never get in a plane with one of those wild women again.[10]

They did have one woman instructor, Helen Duffy, who "came flying into Houston Municipal Airport in late 1942 from Connecticut in her Luscombe. She was a commercial flight instructor looking for a job. Bill Lowrey [one of the instructors] immediately recommended that she be employed to instruct in the women's training program."[11]

We had ground school and flight school there at the airport, but when we went home, we were on our own. We didn't have the strict regulations the girls who came later did.

Photo 10. The early days at Houston Municipal Airport.

Courtesy WASP Archive, Texas Woman's University, Denton, TX.
Marion Mackey, Geri Lamphere, Byrd Granger, Marjorie Kumler, Madga Tacke, and Dorothy Young — all Class 43-1.

There was no such thing as a uniform for the girls of 43-1, but, says Betty Blake, they made the best of it.

We wore khaki pants and white shirts and we had a khaki jumpsuit for flying. I wore cowboy boots for fun. They're perfect 'cause the planes had these two metal strips that your feet were on and the rudder pedals are down there, and that slant of the cowboy boots was the perfect angle to put your feet on the rudders.

Of course we didn't have military planes at first. They didn't think we were going to last even though we had to have two hundred hours flying time to get in. We flew civilian planes painted olive drab with stars on the side. And everyday we flew a different kind of plane.

Lieutenant Fleishman, observing the chaos that reigned across the way, quickly offered to organize a daily physical fitness training session. With Dedie's relieved nod of approval, he began to lead them in daily calisthenics but, as Jane Straughan pointed out, "some of these girls were very temperamental. If they didn't think it was a day to do calisthenics, they sat up on the fence."

He volunteered to be their athletic director and embarked on the nearly impossible. And in order to make them look military, he decided to teach them to march. The free spirits of the Guinea Pigs were tough nuts to crack.

So much for discipline. But they weren't military! These girls, like many of the original WAFS, were older—late 20s and some in their 30s. Many of them—again like the WAFS—were married.

The first military airplanes—Fairchild PT-19s—did not arrive until January 1943. The open cockpit, 175-horsepower, primary trainer (PT), tandem (single front and back seats), low wing, taildragger (the little wheel is in the back), was the same plane the original WAFS began ferrying out of the Fairchild factory at Hagerstown, Maryland, in November 1942. "We never thought anyone was going to let us fly anything like that," says Jane Straughan. "It was a big thrill. And we needed all that cross-country time."

After the PT-19s came the basic trainers—the BT-13 and 15 made by Vultee—and then the advanced trainers: the North American AT-6 Texan and the twin-engine Cessna AT-17 Bobcat.

The first class was expected to graduate in March 1943, but the women couldn't begin to build the required time until the military trainers arrived. When they did arrive, the Guinea Pigs logged those trainer hours with a mixture of pent-up enthusiasm and relief to finally get on with it!

The initial program of instruction, adopted November 10, 1942, called for 115 hours of flying instruction. This was divided into 25 hours in liaison aircraft; 75 in commercial type aircraft (10 transition, 25 instrument, 10 night, and 30 navigation); and 15 in basic and advanced trainers. Twenty hours were to be spent in Link trainer instruction. The academic phase consisted of 180 hours: navigation, weather, aircraft and engines, communications, and procedures of the Air Transport Command. A revision on November 14 called for 315 hours of ground school. Apparently this instruction, at least for the first class, was flexible and was adjusted to the experience level of the trainees.[12]

As the training program got underway, further recruiting convinced Director of Individual Training, Gen. Luke Smith, that there were not five hundred women in the country who had 200 hours flying time. Cochran had already pointed that out to him. With that revelation, the Flying Training Command was authorized to accept women who had 75 hours.

To the 75 would be added 115 hours flown in training—a total of 190 hours. In General Smith's opinion this was close enough to the 200 mark. It was comparable to the Training Command's combat graduates. General George had his doubts based on ATC's prior experience. Right after Pearl Harbor, the Ferry Command—desperate for pilots—dropped its requirement from three hundred to two hundred. The results were bad and the three hundred hours was reinstated. George also suggested that if General Smith's policy was followed, "there would be two standards, one for men and one for women."[13]

Gen. George Stratemeyer, Chief of Air Staff (and General Arnold's right-hand man) put an end to the discussion. "Women pilots who are graduates of the Flying Training Command schools will be accepted by the ATC as qualified for the ferrying of training type aircraft within the continental United States under contact, daylight conditions." "Under contact, daylight conditions" means daytime flying and able to see the ground.[14]

Life in Houston was about to get a bit brighter. In February 1943, the 319th welcomed a new commanding officer, Major Walter W. Farmer. One of the first things the major did was to appoint Fleishman Tactical Officer, Special Services Officer, and Physical Education Director for the WFTDs.[15]

For the women of the 319th, Fleishman was the man who taught them to march. He had them marching to ground school, to the flight line and to meals—singing all the while. For the women of the 319th, he was the man who challenged them to build the muscle they would need to fly ever increasingly heavier airplanes. Daily calisthenics made them use muscles they didn't know they had. They complained, but most of the women came to respect him and apply themselves because of the way he treated them and the challenge he offered.

Major Farmer made other good things happen as well. When 43-1 was due to graduate April 24, 1943, Dedie realized that there were no wings to pin on the white shirts of the twenty-three. Jackie Cochran was ill. What to do? She went to Major Farmer.

He was absolutely the greatest thing that ever happened to us. He was stomping along that gravel path saying, "these gals gotta have wings. They're gonna have wings if I have to tat them myself. Now do something!"[16]

Dedie contacted Cochran's husband, Floyd Odlum, who told her to "dream up something and send me the bill." Lieutenant Fleishman went to Ellington Field next door and bought twenty-three pair of silver AAF Sweetheart wings, slightly smaller than those worn by the male pilots. He and Dedie worked out a design. Fleishman took the wings to a Houston jeweler, who retooled them and engraved the wings using the Deaton-Fleishman design. The triumvirate—Deaton, Fleishman, and Major Farmer—made the wings happen.

Photo 11. Jacqueline Cochran pins wings on Betty Tackaberry at Class 43-1 Graduation, Houston, April 24, 1943.

Photo courtesy Betty Tackaberry Blake's family

Cochran recovered sufficiently to attend the graduation and pin the wings on the graduates.[17] And Rear Admiral Colbert also was in attendance and had the pleasure of pinning pilot wings on his daughter Mary Lou.

Years later, Dedie recalled, "When the gals received their wings, they glowed, yes sir."[18]

The women of 43-1, the Guinea Pigs, were destined to be the second wave of women to fly for the Ferrying Division of the Air Transport Command—the women still referred to it as "the Ferry Command," though the name had been changed. They were training for that job. They were on their way.

Chapter 6

Class 43-2

Fifty-one members of Class 43-2 entered training in Houston on December 13, 1942. They would be the last group to go through all phases of training in Houston. The facility there was proving inadequate and the program was growing. All training would be moving to Avenger Field outside Sweetwater, Texas. When forty-three of those fifty-one graduated May 28, 1943, their ceremony would be held at Avenger. Class 43-2 had many experienced pilots in its ranks, but already Cochran was having to compromise on the hours in order to get a class together.

Two class members, Betty Whitlow (Rummerfield Smith) and Margaret Ann Hamilton (Tunner), had been CPT classmates at the University of Tulsa in 1940. The class limit was twenty—eighteen men and two women—and these two future WFTDs were the two women selected. It wasn't long before Betty and Margaret Ann each bought a one-quarter ownership in a 40-horsepower Piper Cub. And later it was Margaret Ann who told Betty about the program Jackie Cochran was starting.

Quite by accident, I saw an announcement in a magazine—about four lines. Jacqueline Cochran would be interviewing women pilots in Fort Worth. So I got in my little Cub and flew down from Oklahoma right away and saw her. There were qualifications, she said, but I met them. "Be here Monday, we're starting a class in Houston."[1]

I said I can't possibly be here Monday. I have an airplane out there and a job in Tulsa. And a family.

"Well, get it all wrapped up and be back here in two weeks."

So I did. She was very, very businesslike. I think my interview with her was perhaps twenty or thirty minutes. Most of all, she wanted to know about my flying experience. I can see her right now. I remember the positions that we sat in. I was extremely attentive to every word she had to say. You respect anyone who's interviewing you for a position—as a senior and experienced person. I accepted her as that. I never got to know her, though I saw her a number of times in later years—but always at events, in groups. We never would visit.

Now later I got to know Nancy Love quite well. She became a very dear and wonderful friend.

When Margaret Ann got back to Tulsa, she told Betty about the program. By then Betty was out of college and working at Douglas Aircraft inspecting modifications being done on the B-24s brought there.

I made the application and flew the Cub to Tarrant Field, Fort Worth, for my physical. I passed and was assigned to the second class at Houston.[2]

We were supposed to have two hundred hours to be admitted, like the girls in 43-1. I had about one-fifty. And there were other changes. Whereas 43-1 had lived in hotels and rooming houses and provided their own transportation to the airfield, we were lodged in Dedie's tourist courts. We also had this bus. We called it "the cattle car." It was canvas and had benches running lengthwise. It picked us up and took us to the airfield every morning.[3]

Photo 12. Betty Whitlow, Class 43-2, ready to fly a PT-19.

Courtesy WASP Archive, Texas Woman's University, Denton, TX.

We didn't have any real military airplanes at that time. They were using whatever they could get; then they put military designation on them and painted them olive drab. When the first Fairchild PT-19 came in three weeks later—open cockpit, low wing—we were thrilled. A real Army primary trainer. Then came the BT-15s for basic and the AT-6 for advanced.

The weather in Houston could change very fast. I remember the day before Christmas was miserable, cold and wet. Christmas Day was sunny, bright and hot. The changeable weather is probably why they eventually moved the program to Sweetwater. We griped about the weather and having to wait for it to clear in order to fly, but we were having a ball. We were there doing what we wanted to do.

Photo 13. Margaret Ann Hamilton, Class 43-2, climbs into a BT-15.

Courtesy WASP Archive, Texas Woman's University, Denton, TX.

Another pair of friends drove in from Southern California to become part of Class 43-2: Dorothy "Dottie" Nichols and Mary Catharine "Jary" Johnson (McKay), who was interviewed in 1994.

Because there was no flying on the coast after Pearl Harbor, we had to drive inland to Blythe, three hundred miles away, or to Bishop, four hundred miles away, to fly. We were going out there with friends on weekends, but you couldn't get much flying in that way. One night Dottie said, "I'm going to quit my job and go to Texas because I want to get into that training program and I can't do it living here in California." So, I said, O.K. I'll do the same thing.[4]

We quit our jobs, took my car and drove to Fort Worth. First thing, we looked up Jacqueline Cochran's office. We didn't see her, but we were told to go over to the airbase and take the army physical. At that point, we should have known that they were interested in us, because we got to take the official physical that day. Then we got a motel.

I had a little over sixty hours and Dottie had a little over thirty. We were going to fly there in Fort Worth until we got the three hundred hours we thought we had to have — which we figured was going to be a long time. Two weeks later we got a call from Jacqueline Cochran's office. "We have a class starting in December, would you two join it?"

We never knew for sure why they took us. I thought maybe we were guinea pigs to see if women with low time could do it. Maybe they were having trouble getting women to come to the December class because Christmas was a work day. We went to Houston, I think it was the end of November. The first class was already in training.

I think they ran out of women with two hundred hours very quickly. Cochran really wanted to get women into the training, so that's what she had to do. None of us low-time people washed out. And some of the women with more hours did wash out. Apparently it was more a matter of your ability than your time.

The women who trained later at Sweetwater talked about how everything was very strict and they had to keep certain hours. In Houston, things were so different. We lived in this funny little motel, Oleander Court. They couldn't have any kind of military discipline for us living that way. And we had our own cars. We were very relaxed and free, but we were conscientious. We were out at the field every day.

Jackie Cochran came down to visit us—and this really impressed me. When she spoke to us, she was very serious. She said, "Girls, I am going to tell you something and I want you to know this before you get into ferrying. It is not a glamorous life. You won't know from one day to the next where you are going to be sleeping that night." And I thought to myself what else is glamorous but that? I thought that was really strange that she thought that was not glamorous. Of course, that's the way it was. And it was glamorous!

Catherine "Cappy" Vail (Bridge), a CPT program graduate and a member of the University of California Flying Team, was "allowed" to take the secondary ground school course, though not the flying segment.

From my instructor, a bright, enthusiastic young aeronautical engineer, I learned to disassemble an engine. I also learned how to figure out the cost structure of running a Piper Cub, all very helpful to me later on.[5]

I didn't let the fact that I was a woman stop me. I was on the Cal Flying Team. I was the only girl on the flying team, though others were taking CPT. We'd go around to different air shows and things and have competitions with other students. And the boys insisted that I get the jacket,. I said, "oh no, I can't wear the jacket. I'm a girl. That's for the men." And they said, "You're on the team, get the jacket." I still have the Golden Bear insignia logo of Cal, though I long ago lost the jacket. And I'm pictured in the yearbook holding a trophy.

"My parents were putting me through school, so during my senior year at Cal I worked in the office for the airport manager to pay for my flight time. When I graduated in June 1942, I had 75 hours. Then in August 1942,

my mother and I followed my father to Washington D.C. where he was doing war work.

Always my pal, always looking out for me, my father told me he had read about Jacqueline Cochran. "If you're interested, talk to THE person," he said, "don't talk to inferior people in the chain of command." I made an appointment and was allowed to see her in person. I was impressed. Here's a woman with a lot of experience and ability, beautifully dressed—appropriate for that interview. We talked—mostly about me, and that's always fun to do.

At the time, they required two hundred hours flying experience for the girls they were admitting to the first class. I had seventy-five, so I said "I'll get the time, but I'll have to work."

"She said, "Would you like a job working for me?"

Well, I was struck by hero worship—heroine worship. The idea was that I would go with her to Fort Worth, where the Army Flight Training Center was located. At that point, she was recruiting the first class. The women she selected were on trial—how they behaved and how they did the job. She was pleased that I was engaged to Art [AAF 2nd Lieutenant Art Bridge].

There was always an underlying question, can women do this thing? Can they be separate from men without being hassled and without going against the requirements of the Air Forces? Miss Cochran was very careful about that. Behave yourself or you're out. For a lot of us our age, it was the first time we had been away from home.

So I went to Fort Worth. She got me a job in the Flying Training Command Center. I found a rooming house and made some friends. Three of us roomed in this nice home that was taking strangers in. We took the bus to and from our jobs. When Miss Cochran came to town, she had a beautiful suite way up near the top of a luxury hotel—I can't remember the name. Often, when she was gone, she would let us girls go up there and use the suite.

Photo 14. Cappy Vail, Class 43-2, in an AT-6

Courtesy WASP Archive, Texas Woman's University, Denton, TX

I also did some private secretarial work for her, though never at the Training Command Center. I'd go up to her room and she would dictate a letter to me. Using my own shorthand, I could get her words down as she said them, then I'd go home and, using my old college portable typewriter, I'd type about ten copies to make a perfect one. Then she would read the letters, always very carefully. She had no formal training really, and her English was not perfect, so I would make corrections. Sometimes she would say, "I didn't say this" or "I didn't say it this way," but she never picked on the grammar changes.

One time, when she was dictating to me, she said "You know, I envy your education and your whole background."

I thought to myself, my goodness, what she's done without any education, without any family help, and I said to her, "But look at you. You've come up without any help and look where you are." So we had that little sojourn in common, you know, that she wished she had had an education. I liked her very much. I respected her. And she always remembered me, but I wasn't close to her. I preferred to stay away from officials because I didn't want to appear to be buttering them up. There is a Spanish word for it. Vergüezca [bashful, shy]. I know of no word in English like it.

Cappy worked for Cochran from mid-October into December 1942, earning additional flight time in the process. Then in December, Cochran told her that she was to report to Houston with the second WFTD class. "I had no idea how many others, like me, had been admitted, with fewer than two hundred hours. We were asked not to say anything, and we didn't. I learned later that there were several others."

Cappy's roommate in Houston was Marie Muccie (Genaro), who hailed from New Jersey—the other end of the country. Marie's interest in model airplanes led her to find out what real flying was like.

I told the owner of the airport that I would "wash airplanes or direct traffic—anything you need me to do around here." He was so amazed to have a woman wanting to fly, he decided people might be interested to see a woman around. I had a pretty steady job and was building my time—three hundred hours when the war broke out. We lived on the Jersey coast. They closed all those airports on the coast, so I was out of a job. No more flying.[6]

"I wasn't far from earning my instructor's rating. All I had to do was take the test. I checked inland airports. Nothing. But about that time I heard about Jacqueline Cochran, called and made an appointment. I met her in New York.

"I had the required time, but wasn't twenty-one. But by the time they called me, she said okay, just get my father's permission. But he said, "You want me to give my permission for you to go fly Army airplanes and kill yourself?"

That's what I'm trained to do, I told him. That's all I know how to do. But he still wasn't going to give me permission. So I told him I had applied for the Royal Air Force and they didn't have an age requirement. I could get in that without his permission.

"Well, I'd rather have you here in the States,? he said. What he didn't know, and what I didn't tell him, was that the Royal Air Force required pilots to be five foot four, and I was only five two. I wouldn't have gotten in anyway.

Miss Cochran told me that the women's flight training program was an experiment, and that the first few classes would train in Houston, Texas. "Then," she said, "if it all goes well, if they see that you women can fly these Army airplanes just like the men, then we will open up another school in another location."

Photo 15. Six from Class 43-2

Courtesy WASP Archive, Texas Woman's University, Denton, TX.
Standing: Ruth Thompson, Ruth Dailey, Martha Lundy, Marie Muccie; kneeling: Kay Gott and Avanell Pinkely, all Class 43-2, Houston.

On graduation day, we flew all the planes from Houston to Sweetwater—I flew an AT-6—and we graduated there. The only connection I have with Sweetwater is that our graduation was held there.

Ruth Dailey earned her private license January 12, 1939, through the CPT course at Tyler Junior College, Tyler, Texas.

I learned to fly in an old Piper Cub. The guy that taught me how to fly was named Rush Finetrock. One day I said to the airport manager, I wish I had my own airplane. He said, "well, I'll be on the look out for one." So he called me one day and he said, "come up here and I'll fly you over to Kilgore. There's a nice little airplane for sale over there."

Here was this little blue and white Aeronca with wheel instead of stick. $1500. I had to borrow the money. This is Texas, they didn't even lend money on cars in 1939. But I went. I said, Mr. Charley, I need to borrow $1300. He said, "well what are you going to do with it?" I said, buy an airplane. It took me five years to figure out why he let me have it.

He figured I'd kill myself and the family would pay off the plane.[7]

You could go from pasture to pasture all over Texas. There weren't any airports. The first little trip I took in my new plane, I was going to Dallas. Of course, I'd driven it for years. So I headed for Dallas. I looked over and there was a water tower. I thought, I don't remember a little town over there, so I went over to see what was there. I don't remember what it was, but I decided I was off course. I saw another place I didn't recognize from my car travels, so I checked it out. And then I realized I was almost in central Texas. I looked at the gas bobble thing and I thought, it's getting low. So I drew a line to what I thought was the nearest airport. It was Grand Prairie, and I stayed on that line until I got to Grand Prairie.

From then on I was the world's best navigator.

When I met with Jackie Cochran on November 11, 1942, in Fort Worth, she said, "You're just what we're looking for." She told me to take the physical and if I passed I would be in. I did. I passed. Then went home and sold my airplane. I called her and told her I could be in Houston on December 12, and she said, "OK, report for the second class."

Ruth's memories of Houston match those of her classmates. Marching to the flight line after breakfast—the length of the Houston Airport. Morse code was hard for her, but taking engines apart and putting them back together was not. Nor was weather, as she already had a native Texan's gut knowledge of what caused the weather to be what it was and she had been flying in it long enough to put two and two together.

I was scared all the time that I was going to wash out. I think our biggest disappointment was, when the first class graduated we found out we weren't going to be commissioned. When we first entered training at Houston, we raised our hands and took the officer's oath. We all assumed we would be second lieutenants like the guys. But no. The Secretary of State said that the law says that "when HE has completed his training, HE will be commissioned." So that's history for you."

Martha Wagenseil (Davis Rupley) had 129:15 hours on November 9, 1942, when she and two non-flying friends left Pittsburgh Butler airport in her three-seat, tomato-red Piper Super Cruiser—West Coast-bound. She needed 70:85 hours to qualify for the Women's Flying Training Detachment. She had an appointment in Fort Worth for an interview en route home.[8]

I had to have the time and I had to have it in a hurry, so I resigned my job, sent to the U.S. Coast Guard Geodetic Survey for maps of the United States, convinced two friends who knew nothing about flying to go with me to hold the maps, and got ready to go.

Flying on the West Coast was closed to civilian traffic, so we landed in Phoenix, left the airplane there, and drove west in order to see the blue Pacific. On the way back east, I stopped in Fort Worth for my WFTD program interview, then I went home to wait. There was nothing I wanted to do with my whole life right then but fly. Early in December, the telegram came.

A week later, I was on my way, driving southwest out of Pennsylvania. On December 15, I reported to Houston AAB at Houston Municipal Airport. There, I met my classmates.

The members of the first class were living around the city in rented rooms. We, 43-2, had an entire motel. It was clean, it was nice, it was comfortable. We had a valet and maid service, we had clean linens, we had no complaints about our quarters, none at all.

Iris Cummings (Critchell) graduated from the University of Southern California in 1941 with degrees in physics and math. In addition, she had earned her private pilot's license through the CPT program at USC and had all her ground instructor ratings. In September that year, Iris was asked to teach all the ground school classes in the CAA's advanced Civilian Pilot Training Program at Brackett Field in Laverne, California. While teaching, she earned more flying time on Wacos and Stinsons in cross-country and instructor programs.[9] Then on December 7, with the news of Pearl Harbor, all the airplanes were grounded. The flight schools on the West Coast all moved inland where those students finished their training.

Some of my former students were in an advanced course in Carson City, Nevada, and found that they needed an instructor up there. They suggested me. I signed on for what turned out to be three, 16-week semesters teaching Navy cadets before they went on to Pensacola. It was a good preflight program. I liked what I was doing.

My third semester didn't end until December 1, 1942. By then, the WAFS already were flying out of Wilmington, so I opted to join Jackie Cochran's Class 43-2 at Houston.

I began my Army flying on December 18, 1942, in a Piper J-3 Cub. On January 11, I had my first flight in a Fairchild PT-19. That was the east coast's military primary trainer, built in Hagerstown, Maryland. Its mid-western and Texas equivalent was the Stearman PT-17 built in Wichita, Kansas. The Ryan PT-22 was the primary trainer used on the West Coast.

But what they brought to us in Houston was the PT-19 Fairchild—the low wing, 175-horsepower Ranger engine, open cockpit, fixed-gear trainer. From then on, my logbook shows about fifteen flights in the PT in January, and I was already doing cross-countries Houston to El Paso.

On January 31, I got my introduction to the Link trainer.

I began the second phase of training in the BT-15 on February 24, and a month later, logged my first under-the-hood instrument training time, also in the BT-15. We did hood training—instrument training—in the back seat of the BTs.

I took my first AT-6 flight on May 1. By then they were hurrying up to get us finished. Quite a few of the AT-6 flights were at night and I was logging pilot-in-command hours on cross-countries.

On May 15, I was checked out in the twin-engine AT-17D—also called the UC-78—the Cessna Bobcat, the plywood airplane. By the time May 26, graduation day, arrived, I had logged twelve flights in the AT-17.

Class 43-2's graduation ceremony was to be held at Avenger Field in Sweetwater. And since all training was being moved to Avenger, as of that date, the graduates as well as 43-3 trainees got to fly the remaining aircraft up to the school in Sweetwater. I flew an AT-17 from Houston to Sweetwater and transferred the airplane to the supplier of aircraft equipment at Sweetwater. Two classmates flew with me as passengers.

We left Houston very early in the morning, so we were up at Sweetwater early in the day—in plenty of time for the graduation ceremonies. Then those of us who were ready to go out on the train were taken right to the railroad station. And we were gone. We were there at Sweetwater only long enough to park the airplanes, go through the graduation ceremony—which was much shorter than the previous one for Class One—and we were off on the train that afternoon.

And thus ended the training phase for Class 43-2.

Chapter 7

Class 43-3

Class 43-3 began training January 16, 1943. Fifty-five reported and they were the last class in which all of the trainees reported to Houston. In February 1943, half of 43-4 would report to Houston and the other half to Sweetwater where, after May 18, all WFTD training would take place. Forty-three members of 43-3 would graduate July 3 in Sweetwater.

Female aeronautical engineering graduates were a rarity in 1941, but Purdue University claimed one in Lois Hollingsworth (Ziler). A top scholar, she went to work in the research division of United Aircraft (later United Technologies) in Hartford, Connecticut. United was associated with such prestigious names as Pratt and Whitney engines, Sikorsky aircraft, and Hamilton standard propellers.[1]

The summer after I graduated from high school, I started my flying. That was in 1937. I had soloed and logged quite a few hours before I went to Purdue. My father had left us a trust fund, though it was greatly reduced due to the Depression. Still, there was approximately sixty dollars each for my sister and me every month. It's amazing what sixty dollars would do in those days. It paid for my flying.

I read about the women's flying program in the paper and through the Ninety-Nines group that I had joined. So I wrote and got papers and was told to report to Jacqueline Cochran in New York. I had a little over two hundred hours then. But I didn't have my commercial license because the summer before, they had stopped civilian flying on the east coast.

On the appointed day I went to New York City for the first time. It was a sixty-mile trip by train. Cochran met us and then we conducted our interviews with Ethel Sheehy. She filled in the blanks. A few days later I got notice to report to Windsor Locks, Connecticut, Army Air Base for my physical. I just laughed. I thought it wouldn't do me any good because I wore glasses and had all my life. I knew I wouldn't be accepted. But I was —with my glasses. Every time I had to change bases, I had to take another physical and I just knew they would find out that I couldn't see without my glasses and send me home. For some reason or other it didn't seem to matter. I was told to report to Houston, Texas, on January 15, 1943.

We gathered and were told what we were in for and were told if any of us wanted to go home we could. We were civilians, but we would be under military control. We didn't have very good facilities, so if anybody wanted to go home she could. I don't think anybody wanted to go home. We were there. They didn't have any housing facilities. They housed different groups at various motels. I was at the Alamo Motel. The one next door was the Oleander Courts.

We began training January 16. I ended up helping quite a few of the others that were having trouble with the aerodynamics, the navigation, and engines, but I don't know of anyone who did not pass ground school eventually. Ground school included Morse code and that was hard for me. I had done it in Civil Air Patrol, but still it was hard to get the speed up.

We were split up in two flights. Half flew in the morning while the other half of the class went to ground school, and then we would switch. You got to know the girls in your flight pretty well, but you might not see the other girls unless they happened to be in your motel.

We were divided into marching platoons. At breakfast we would form up and march to the flight line, and then march from the flight line down to this place where we ate lunch. I enjoyed marching—thought that was fun. We got pretty sharp at it. It's a way to move a lot of people at a very efficient pace to an objective.

Lois Brooks (Hailey) was introduced to flying when her father took his children for rides when the barnstormers came to town. But it was her stepbrother, Jim, and his friend, Vic, who really got her in the air.

One flight sold me—up there close to the clouds and looking out over the world. It was really great. My first instruction was in an old 40-horsepower Taylorcraft. Then Jim and Vic and I decided to buy a new Taylorcraft. It was a 1939 model, single ignition, no brakes and a tailskid—$1995 at the factory. We each put a dollar an hour into our central fund and that would pay for the gas and oil and inspections. We'd have a little bit left over so then we bought brakes and the tail wheel—improved it a little bit.[2]

Both Vic and Jim got their commercial licenses within a year and they were ready to sell out. I bought out their interest for $333 to each one of them and I took the airplane out to Minden [Nevada] *where I was teaching school. Some of the schoolboys helped me drag a landing strip on steer grazing land. Sometimes we had to drive the cattle off before we could take off and land.*

When Pearl Harbor came along, all of the airplanes were grounded. I had to move mine over to Carson City so they could check me in and out when I went up. Of course, nobody could even fly on the Pacific Coast. A lot of people came up to the Reno-Carson area to do their flying.

I got a telegram from Nancy Love saying that if I would like to be a part of the WAFS, I could come back to Wilmington, Delaware, and take a flight test. Well, by then I had a commercial license and 500 hours, but I thought, gee, I might go back there and flunk that test, so I was afraid to go. Besides, I really liked the job I had teaching in Minden. I had an elementary school band. I was so proud of those kids, I didn't want to leave them. So

I didn't go. Later I got a telegram from Jacqueline Cochran to meet her in San Francisco for an interview.

Well, I still didn't plan to go, but I wanted to meet Jackie Cochran so I made the trip down to San Francisco. She wasn't there, but Ethel Sheehy was. I told Ethel that I had no intention of joining the military, I just came to meet Jackie. She said "Well, if you should change your mind you can go to Mather Field [Sacramento] *and take the physical."*

I had heard all the guys talk about this Six-Four physical exam, which was tough.[3] *I knew I couldn't pass it because I wore glasses, but I thought, well, I'll go down just out of curiosity and see how it is. I went down with Iris Cummings. They put me through the physical and they passed me. Iris went into the second class but I didn't go until the third class, after Christmas vacation—mid-semester.*

We never talked about hours. I don't even know if our instructors knew how many hours we had. I don't think they did. It wasn't discussed and for me to brag that I had a commercial license wouldn't have set well.

Our training at Houston was uneventful except that one girl was killed in a PT-19 with her instructor.[4] *They came back and told us it was pilot error. And I remember thinking, how could it be pilot error if there was an instructor in the airplane? But as time went along, I would see a lot of pilot error.*

Lillian Connor (Roberts-Risdon) came to the Class of 43-3 from yet a different background.

Photo 16. Lillian Conner, Class 43-3.

Courtesy WASP Archive, Texas Woman's University, Denton, TX.

General Luke Smith, Army Flying Training Command, appointed Connor to serve as Jacqueline Cochran's assistant after General Arnold gave Cochran the go-ahead in September to establish a flight training school for women. Cochran knew nothing of Connor's appointment, whose title, in Connor's own words, was "Assistant Director of Travel Operations for Female Pilots or something like that."[5]

I don't know when or how he told her that I was coming. All I know is, when I arrived, I was simply dumped into this office with my desk next to hers. I did not know that she hated me before I even got there. That was because she was being told to do something and, with Cochran, that was not the way to go. Not if you wanted to be popular and within her good graces.

Anything that I took on would be something that she didn't want to do anyway. What I wound up doing quite a lot of was going on inspection tours of various fields and bases. My presumed reason for doing this was that I was to see if they were using women to do much. If they were, I was to congratulate them. If they weren't, I was to suggest more ways for them to use women. Sometimes if Cochran wasn't around, I would have to interview the people that wanted to go to the flight training school.

Initially, the Army Air Forces Flying Training Command[6] was located in the Old Munitions Building in Washington, because construction of the Pentagon was still underway.[7] Those offices of the Flying Training Command were moved to the Pentagon in fall 1942. Cochran, however, was moved to the offices in Fort Worth, Texas. There, the Flying Training Command occupied the top four floors of the Texas and Pacific Railway office building.[8] It made sense, Connor says. "Texas was just loaded with Army Air installations."

In Washington, Cochran was in everybody's hair. Nothing could have worked out better than to get rid of her by putting her in Fort Worth instead of right next door where she could walk right into anybody's office anytime and make trouble.

I didn't go with Cochran to Fort Worth because I wanted to attend the flight training school. I wasn't in the first or the second class because I had to wait for her to get out and get all her stuff moved down to Texas. Then I had to get everything set up in the new quarters in the Pentagon and finish up any appointments that I needed to take care of before going down. I couldn't get everything all tied up with a nice pink ribbon for the office in Washington until it was time for the third class.

At the training school, and in ferrying later on, there was no such thing as a Sunday day of rest or a Saturday away from your job. It was a seven-day-a-week situation. You did the same thing seven days a week. You never slept late. You never went off and did some snarky thing someplace. You just stayed in school, did all the flying, got all the hours that were required of you, took all of the ground school that was required. Passed it all, because not passing could hold you up.

They told us that our ground school was the equivalent of two years of college—the subjects covered and the level at which we became proficient in those subjects. All of this went on your qualification sheet.

None of it was particularly easy. You had to learn an awful lot of math —and geometry came in handy. If you ever got the isosceles triangle in your head, you had a different concept about flying than you would have without all that. It made you aware of all the various stresses, strains, pulls, and pushes that are going on when you're in an airplane off the ground. And I can easily see why that was required. And why it would make you eligible to say that you had a couple of years in college.

In training, the only thing we ever wore was the zoot suit [flight coveralls]. *You see we were never off the base. You have to keep stuff clean. Airplanes are not clean. They're dirty. That's the nature of an airplane.*

At that time the men's uniforms included pinks and greens. Usually they mixed pinks and greens. But as far as I was concerned the greens were the smart move. Because the greens didn't show dirt—very dark. I got a couple of everything in green—the shirts and the trousers. So whatever got too smelly first would get laundered while you were wearing the other.

***

One of Dedie Deaton's rites of passage as the woman-in-charge in Houston came about because of Lillian Connor's connection to Gen. Luke Smith. Deaton told of this connection in one of her oral histories.[9]

Lillian Connor was the gal who came to audit the course, not to be subject to the rules. She was from Washington, and was Gen. Luke Smith's secretary. She said she was there to audit the course and inspect the training. I didn't know anything about it. Jackie knew nothing about it.

I said, "Mrs. Connor, I'm very sorry but you're either an enrolled trainee, billeted where the rest of the trainees are, subject to the regulations that the trainees are subject to, or you may go back to Washington."

"Well," she said, "I'm here under order from General Luke Smith."

I said, "but I am under order from General H. H. Arnold."

We enrolled her and we made her do like the rest of them.

When General Smith came to Houston, he told me that he wanted to take his secretary, Mrs. Connor, to lunch. "I'm picking her up at eleven-thirty," he said.

I said, "I'm sorry General Smith, that is not permitted."

And he said, "but she's my secretary and she's here on my orders."

And I said, "Yes, but she's under General Hap Arnold's orders when she's here. And she is flying at one o'clock and has ground school until noon."

"This is most unreasonable," he said, "I'm sure you can make an exception."

And I said, "General Smith, there are no exceptions on this base. Rules are rules and they are made to be kept. We have a great many young women here who have very important relatives—senators, governors, and generals. And we don't care if they are nursemaids or waitresses or doctors, they are subject to regulations."

Well, he just liked to have died. So he said, "I assume I can take her to dinner?"

And I said. "You certainly cannot. You're not related to her."

And he didn't.

Dedie didn't limit her iron-hand-in-a-velvet-glove treatment to visiting generals. When base commanders or their underlings tried to undermine her authority, she held the trump card.

I had absolute authority over everybody—including the military—when it came to the gals. I had instructions to call Jackie Cochran or if I couldn't get her, or preferred to, I could call General Arnold any time I wanted to.

"You're getting a little bit too big for your britches," one lieutenant said to me. "I am Commandant of cadets. I have authority over what the girls do." [This was later, in Sweetwater.] *Then he told me I'd be off the base in a week for defying him.*

"I don't think I'll be the one that's leaving, " I said. He was sent to the Aleutians.[10]

On May 18, as 43-3 was finishing up basic training, they were sent to Sweetwater flying the BTs—not to return. They moved in tandem with the members of Class 43-2 who were flying the AT-6s and AT-17s to Sweetwater to take part in their graduation ceremony. There were two girls to a plane. Some of them had cars to drive and others went by public transportation.

There were overriding reasons for the move to Sweetwater, in Lois Ziler's opinion.

The lack of facilities at Houston! There still were no barracks. Sweetwater had barracks, hangars, and the necessary set-up already in place. Also, there was the weather.[11]

We had pretty lousy weather in Houston. Some days we could not fly at all. Some mornings we had to wait until nine or ten for the fog to lift. Then the lessons were shortened or some of us didn't even get in the air. Sometimes there would be thunderstorms when the afternoon flight needed to fly. We got our flying time in, but we were behind.

So we moved to Sweetwater, got there nice and safe, and the next day it started raining—for two weeks! That really put us behind. But we were able to finish up all of our ground school so that when the weather improved, we would fly every airplane that was available. We also flew some Saturdays and Sundays trying to catch up. We had our night flying to get into when we went to the advanced airplanes—the AT-6 s and the twin-engine AT-17s. Some of us were flying in the morning, again at night, and then the next afternoon. So we did catch up and graduated practically on time—July 3.

Chapter 8

Classes 43-4, 43-5, 43-6

The Last of the Ferry Pilots

Classes 43-4, 43-5, and 43-6 were large classes—152, 125, and 123 respectively. The previous three classes had begun training with 29, 51, and 55 respectively. All graduates from the first three classes were sent directly to the Ferrying Division. But when Class 43-4 graduated August 7, 1943, twenty-six of its graduates were immediately assigned to a non-ferrying base. Some graduates from Classes 43-5 and 43-6 would be sent to non-ferrying jobs as well.

The women of 43-6 were the last class from which the bulk of the graduates were sent to the Ferrying Division. No women from 43-7 were sent, directly, to a ferrying squadron. By the time Class 43-6 graduated on October 4, 1943, many changes had taken place in the Ferrying Division as well as in the women's flying program. That is the story that unfolds throughout the remainder of this book.

Half of Class 43-4 reported to Houston and the other half to Avenger Field outside Sweetwater, about forty miles west of Abilene. They arrived in mid-February 1943. Life at Sweetwater would prove to be very different from life in Houston.

JEAN LANDIS, 43-4

Jean Landis was dressed in her cap and gown rehearsing for graduation from San Diego State Teachers College in June 1940 when she heard, over the loudspeaker in the gym, the announcement that a CPT class would be offered to the first twenty students to sign up at the dean's office. [1]

I could hardly wait until rehearsal was over. At the word "dismissed" I ran out the back of the gym. I took a short cut down a canyon, jumping sagebrush, boulders and cactus—my cap and gown flapping in the wind—sprinting all the way. I arrived at the dean's office, somewhat tattered and torn. I was the first one to sign up.

"Pearl Harbor changed our lives forever," says Jean, who by December 1941 was teaching Phys Ed at her alma mater, Grossmont High School in El Cajon, California. A year later, February 1943, she was on her way to Houston.[2]

There were no barracks available, no quarters of any kind. We lived in third-rate motels. They sent open-bed trucks through the outskirts of the city to round up their student pilots and get us to the field every morning. We waited—in the early morning darkness, cold and hungry—for the truck to take us for breakfast at Municipal. This was the start of each and every day. Days filled with flying, learning, excitement, anticipation, worries and hopes and dreams of graduation day and silver wings. And finally, flying for our country. I never will forget the emotion we all felt on April 24, 1943, when class 43-1 graduated and received their silver wings. The band, the speeches, the colors, the twenty-three marching so proudly after six months of vigorous training. It was so inspiring. I especially remember little Dottie Young, barely five feet tall. We called her the little colonel. What a gal.

Photo 17. Jean Landis, Class 43-4, on the wing of a PT-19.

Photo from Jean Landis' personal collection

Houston had reached its capacity. Three months after I arrived there, we relocated to Sweetwater to complete our training. As we flew our planes into

Avenger Field, the Army cadets moved out. The base was almost transformed over night from a bunch of guys to a bunch of gals.

The move to Sweetwater brought order and discipline not imposed on the women in Houston. In Sweetwater, the women lived under a demerit system and were organized into squadrons. Jean became a Squadron Commander. Her job was to march the squadron to meals in a military fashion.

Nothing else changed. We were issued the same army zoot suits, some still hot and sweaty, and always two sizes too large. We had the same Army instructors, army PT, army food, army discipline, army barracks, army orders, and the dreaded check rides.

The terror of Army check rides became a universal fear among all of us, even those who considered themselves hotshots admitted they had succumbed to that fearful state of mind. How could you keep from it after seeing a number of your friends and classmates pack their bags and leave with tears shed by all of us?

Finally, the long-awaited day—August 7, 1943, graduation day—arrived. After six months of training, we had earned our beloved silver wings.

Dorothy Kocher, 43-4

Dorothy Kocher (Olsen) of Woodburn, Oregon, dreamed big.[3]

I wanted to fly airplanes from the time I was eight years old. I read Red Knight of Germany. I wanted to fly military airplanes, and I wanted to fly fighters!

When I first went to Houston I hated it, and if it hadn't been for losing face, I would have come home the second day.

Then I received word that my fiancé had died of spinal meningitis, contracted in an epidemic at the camp where he was stationed. I went home for his funeral. When I came back, I was way behind my class. They gave me a check ride to see whether I could stay with my class or whether I had to fall back a class.

I had the worst cold I'd ever had in my life. I was so sick. But my check ride pilot was wonderful. We went up and we did aerobatics. I passed. I hung with my class, but I had to do an awful lot of flying to catch up. I was so lucky, 'cause I wouldn't have gotten Long Beach if I'd been held back. Long Beach was the base that was the key to the fighters. I would have been heartbroken if I hadn't gotten Long Beach.

No graduates from the next class, 43-5, were sent to Long Beach.

Virgie Jowell, 43-4

Virginia Lee "Virgie" Jowell (Hagerstrom) was working as a reservationist for Delta Airlines in Dallas when she heard about the training program.[4]

We needed a hundred hours at that time and I only had seventy-five. I bought the time at Dallas Aviation School because it was near where I lived. Then, to give me flying time, I took the graveyard shift because nobody else wanted it. I would go out to a small auxiliary field—we had to go by bus—and maybe I'd get in only one hour flying, but I'd hang out there all day. Then I'd come home, grab a little sleep, shower, and get back at midnight to Delta Airlines. There was no business that late, so I could sort of shut off the lights, put my head down and sleep a little bit. We only had one plane coming in at 3 a.m.—I flew the last hour right before I had to get on the train and go down to Sweetwater.

There, I ran into future classmates Grace Clark (Fender) and A.J. May (Starr). They took me up to their hotel room at the Bluebonnet because I had no reservations. We talked all night.

Photo 18. Virgie Jowell, Class 43-4.

Courtesy WASP Archive, Texas Woman's University, Denton, TX.

A.J. May, 43-4

Alice Jean "A.J." May (Starr) from Norwood, New Jersey, took her first flight lesson in a silver all-metal Luscombe on floats. It was 1940 and an organization called the Women Flyers of America (WFA) was recruiting women to learn to fly to help with America's defense if war came. [5]

I signed up immediately to fulfill a lifelong ambition! I was nineteen years old, an art student in New York City, with various part-time jobs to pay for tuition, transportation, and supplies. Getting to Skyharbor seaplane base on the Hackensack River for flying lessons involved riding three buses plus a two-mile walk—all of which took more than two hours.

I heard about Nancy Love and the WAFS in Wilmington, but didn't have enough hours. I went to work as a Link trainer instructor at the Naval Air Station in Atlanta. That's where I was when Cochran's representative Ethel Sheehy sent me a wire to report for the training program in Sweetwater."[6]

I had 82:55 hours. My claim to fame is that I was the first bugler at Avenger Field." [This was before Nelle Carmody arrived with Class 43-6.] *I had played trumpet in high school, so I blew "Taps."*[7]

Vi Thurn, 43-4

If anyone from Class 43-4 could lay claim to pioneer roots, it was Violet "Vi" Thurn (Cowden). She was born in a sod house on a farm in Bowdie, South Dakota.[8]

I would watch the hawks fly and zoom down on a chicken with such speed and accuracy. I thought if only I could soar up in the air and zoom down like that, what fun that would be. The dream of flying was planted in my mind at that time.

I got my chance to learn to fly while I was teaching in Spearfish. The airport was six miles from where I was living. I didn't have a car, so I rode my bicycle out early in the morning, took my lessons, then rode six miles back and taught my classes. I went to ground school at night. I loved it.

I had seventy hours. It was a dream come true. Twenty-seven years old and here it was happening. I was on my way to Sweetwater. It's like Disney says: "If you can dream it, you can do it."

The day that I graduated, General Tunner came in. He wanted to know if the girls could really fly. Lucky me, I had to ride with him the morning I was supposed to get my wings. I don't know how I was selected. I don't know whether any others had a ride either. He had me do a few maneuvers, but nothing that was impossible. I was comfortable with it.

Later that day, my new wings pinned to my blouse, I put my hand over them and vowed that no one was going to take those wings away now.

She was on her way to Dallas and General Tunner's Ferry Command.

Class 43-5 reported to Sweetwater in late March 1943. This was the first class to be totally trained at Avenger Field.

Gail Bevis, 45-5

Photo 19. Gayle Bevis, Class 43-5.

Courtesy WASP Archive, Texas Woman's University, Denton, TX.

Kansas City-born Gail Bevis (Reed Ewing) had not always dreamed of flying.[9]

It never crossed my mind. My motivation was the chance to win a CPT scholarship. I just wanted to go back to school and show them how smart I was.

I didn't win. I wasn't as smart as I thought. They gave the program again. The second time, I made the top ten and won the flight scholarship. My first lesson—the first time the wheels left the ground—well, my life has never been the same! I was hooked."

The Ninety-Nines and working as a Link Trainer instructor led me to the WFTD. I had just under the required seventy-five hours when I was accepted into 43-5. I was twenty-five the day I left home—March 23, 1943. There I was on a slow train to Texas, dressed in shoes and stockings with a seam up the back of the leg, a suit, hat, gloves and bags. The hat probably had a veil. I was on my way to Sweetwater with all the other girls.

In Sweetwater there were six of us and a bathroom in between and six more and a partition and so on. In the bathroom, there were two johns, two basins, two showers—six girls on either side—twelve. It was no problem. This is 1943. Houses usually only had one bathroom.

We were cut from a different bolt of cloth. But these girls became your friends for life. Flying was a mindset."

Harle W. Pinkard was my instructor. He was tall, kind of normal-looking guy—big nose, kind of a shock of kind of sandy hair and cowboy boots. I thought, "Oh boy." [Laughter] *Cowboy boots were not chic in 1943. He was wonderful! We had a rapport. He taught me everything I know. I turned out to be an average WFTD trainee, but I was good when I was with him or felt that I was.*

He would talk from the backseat and say things like, "Ata-doing it." Or "How about giving a little dual on those Chandelles?" If I wasn't doing so well, he'd say, "Eh, you can get away with that now. This airplane will take care of you. But not when you're out there flying Mustangs."

Sylvia Dahmes, 43-5

Sylvia Dahmes (Clayton) from Redwood Falls, Minnesota, followed her brother to Western Union College in LeMars, Iowa, and enrolled in CPT there. Since they allowed one girl for every nine boys, her brother thought she should be the girl in the class. Both applied, but her brother turned out to be colorblind and was disqualified. Sylvia, on the other hand, was accepted. [10]

I wasn't going to do it if he wasn't in the class. He was my protector. But he said, "You spent your five dollars for the physical, you've got to stay in." The boys—football players, basketball players—were the popular ones and I was scared. But they were all good to me. I took my CPT in the fall of '40. I was the only girl in the school who could wear slacks to the dining room or to classes. I got my license in January of '41.

Two years later, I was working as a medical technician in a Detroit hospital. A nurse, who flew on the transport planes bringing wounded soldiers back, told me to go out to a nearby airport where the Civil Air Patrol was operating. "They get gas for their planes and they practice flying search and rescue missions," she said. "You get to study the rules and regulations, aircraft engines, weather."

I went and liked it because, once again, I could talk pilot talk and keep up with pilot news. That's where I learned about Jacqueline Cochran and the program. I went in with 65 hours. They were asking us to have 75. Then I had the interview, and they called me and said they would accept me without getting the last ten.

Class 43-6 reported to Sweetwater April 25, 1943.

Adeline Wolak, 43-6

Photo 20. Adeline Wolak, Class 43-6.

Courtesy WASP Archive, Texas Woman's University, Denton, TX.

"My father flew. He loved Cubs!" says Adeline Wolak (Ellison), who grew up in Chicago. He wanted Adeline to learn to fly.[11]

Every Sunday he took me to the airport for lessons. Finally I got my eight hours, then I soloed. I remember my instructor telling me, "When you take off, this plane is going to go up like a homesick angel."

I took off. Oh god, I used to love to do loops and spins and this is with eight hours training. I was up there having a ball. But when I landed, he was really mad. "For crying out loud, I thought you were going to spin right into the ground."

My dad and I both joined Civil Air Patrol after Pearl Harbor. I really looked forward to our meetings. CAP is where I got my first taste of cross-country flying.

I saw an ad in the paper seeking women for pilot training. I took all the tests and was accepted into the program, Class 43-6. A few months later, my mom and dad were borrowing on their insurance policy to make the long train trip to Sweetwater for graduation. How proud they both were![12]

Anna Flynn, 43-6

Photo 21. Anna Flynn, Class 43-6

Courtesy WASP Archive, Texas Woman's University, Denton, TX.

Anna Flynn (Monkiewicz) from Natick, Massachusetts, was one of several WFTDs who were "Piper Girls." They worked for William T. Piper, Sr., building Cubs in Lock Haven, Pennsylvania. She worked in the aileron department.[13]

My job was to put rivets in. That was my job—rivet, rivet, rivet. These weren't the kind of rivets that required a machine. They were soft aluminum ones. You just tapped them with a little hammer and they flattened out. They didn't want the wing to weigh too much by putting steel rivets in it.

Everyone who worked for Piper had the opportunity to learn to fly and I took full advantage. Most of the time my mind was on flying.

In the fall of 1942, I was installing rivets, when here comes this big transport and all these uniformed girls got out. It was the original WAFS. They came to ferry Cubs. They looked like movie stars to us. They took over the delivery of the military Cubs at that point. [Note: the actual date was October 22, 1942.]

Not long afterwards, several of us Piper women received a wire from Ethel Sheehy, Jackie Cochran's recruiter. Mrs. Sheehy asked for anyone interested to come to State College. She talked to us and a lot of us signed up and went and got our physicals. Then we just waited for a wire that said come to Sweetwater.

I rode the dirty old anthracite coal train—full of GIs. April 25, 1943, is when my class got in. Those planes just seemed huge compared to the Piper.

There were about twenty girls from Piper Aircraft, and quite a few of us in 43-6. Mr. Piper came to our graduation and pinned our wings on.

Enid Clifford Fisher, 43-6

When Enid Clifford started the CPT course, she was working as a model and hostess in the gown shop of a large Portland, Oregon, department store. She needed to ask permission to be a few minutes late arriving for work in order to get back from Swan Island Airport. When her boss found out why, he was so impressed he had her pick out a pair of slacks from the sportswear department as a gift. It was her first pair of slacks.

Enid married Donald Fisher on October 17, 1941. Early in 1943, she became a recruiter for Jackie Cochran, interviewing women applicants for the flight-training program. Once she was sure that Donald, an ensign in the Navy Engineer Corps, was being sent overseas, she applied and was accepted for Class 43-6.[14]

Dorothy Webb, 43-6

Dorothy Webb of San Jose, California, earned her private pilot's license in 1935, the beginning of a flying career that eventually grew to 2,500 hours. She was in the air with a radio on December 7, 1941, when the attack on Pearl Harbor occurred. When she landed, she and four other women who were flying that day wrote to Washington, offering to fly light aircraft in the war effort. Fifteen months later, Dorothy entered training at Sweetwater as a member of Class 43-5.

Dorothy was married to a B-25 pilot instructor at Mather Field, near Sacramento. He was killed while on a night training mission. The bomber crashed near March Field, Riverside, killing all aboard. At that point, Dorothy was in the middle of her training at Sweetwater. She went home to Sacramento for the funeral, after which she saw to her young daughter's continuing care with her parents in Bakersfield. Then Dorothy returned to complete her training with the next class, 43-6. When she graduated, Dedie sent her to Long Beach so that she could be near her daughter and her family.[15]

All the women included in this chapter were sent to the Ferrying Division and went on to become pursuit pilots in 1944. They qualified to ferry the Army's fastest aircraft, the single engine P-51s, P-47s, P-39s, P-40s, and P-63s from coast to coast.

(Thurn's, Fisher's and Young's information did not come from an oral history.)

The graduates from Houston and later Sweetwater didn't know it, but Dedie Deaton held the key to their assignments after training. [16]

I didn't believe in drawing names out of a hat. I felt people ought to have the privilege of going where they wanted. I made out a questionnaire: where they wanted to go, the aircraft they wanted to fly, the people they wanted to be assigned with. Then I worked out a schedule per city and I got ratings from the flight line, the ground crew and ratings from their student officers. I made ratings from the demerits. Then I figured out who had the highest rating and made a list of them for different choices. Of course their rating on the flight line had an influence on whether they went to those places with the equipment they wanted to fly. Then I assigned bases giving the first choice to the people who had the highest average.

Dedie Deaton earned her Texas-sized spurs as the woman-in-charge of the WFTD trainees' welfare first in Houston and later in Sweetwater. Most of the trainees, when asked years later about her and her influence, wax eloquent and long and with great feeling. A few—there are always a few—ran afoul of her, felt the sting of her ever-increasing capability of handling matters, and feel differently. *C'est la vie!*

Part III

Three More Women's Ferrying Squadrons Formed

January–June 1943

Two of Nancy Love's new ferrying squadrons—one located at Dallas Love Field, Texas, and the other in Romulus, Michigan, outside Detroit —were scheduled to launch in January 1943. The third, in Long Beach, California, would start up in February.

As of March 24, 1943, the original twenty-eight WAFS had completed one hundred ninety-four deliveries.[1] They had been ferrying aircraft for five months.

From January to June 1943, the four women's ferrying squadrons functioned under Nancy Love's exclusive leadership. By early June, with the addition of the graduates of the first two WFTD classes, the strength of each of the four ferrying squadrons stood at roughly twenty. This meant that eighty women were now ferrying trainer aircraft for the Ferrying Division, Air Transport Command.

The following chapters tell the stories of the growth of the four women's ferrying squadrons.

Photo 22. Dallas, three Original WAFS in winter flight gear: Dorothy Scott, Betsy Ferguson, and Florene Miller.

Courtesy WASP Archive, Texas Woman's University, Denton, TX.

Chapter 9

5th Ferrying Group, Love Field, Dallas, Texas

As January 1943 dawned, WAFS Nancy Love, Florene Miller, and Helen Richards flew to Dallas. Betsy Ferguson and Dorothy Scott drove Dorothy's car cross-country by way of Betsy's hometown, Coffeyville, Kansas, and from there continued south to their new duty station, the 5th Ferrying Group at Love Field in Dallas.

During World War I, the Dallas Chamber of Commerce negotiated with the Army Signal Corps (precursor to the Army Air Corps) to establish an Aviation Training School in the Dallas area. The Army opened Love Field seven miles north of downtown Dallas on the south side of Lake Bachman on October 19, 1917.[1]

When World War II came along, Dallas was the perfect locale for a military flight base. It was central to numerous aviation schools, manufacturers, and modification centers. The Army reactivated Love Field as a major military aircraft modification center and made it the home of the 5th Ferrying Group: the men, and later the women, who would ferry those aircraft. The women would deliver to all points across the United

States. The men would deliver to all points across the globe. Aircraft built in the Dallas-Fort Worth area, and ferried by the men of the 5th Ferrying Group, included the B-24 Liberator, built at Consolidated's Fort Worth plant; the AT-6 trainer and later the P-51 Mustang, manufactured at the North American plant at Grand Prairie; and the P-38 Lightning, which was modified at Love Field's Lockheed Mod Center.[2]

When the first five women ferry pilots arrived on the scene in Dallas, they began to transition into and ferry basic trainers—BT-13s and 15s. Then in the spring of 1943, they were sent to the Boeing plant in Wichita, Kansas, to pick up and deliver PT-17s—the same primary trainers six of the WAFS had flown from Great Falls, Montana, to Jackson, Tennessee, in December 1942. Later on, Wichita was their destination to pick up single- and twin-engine Cessnas and Beeches manufactured there.[3]

Florene Miller soloed at nineteen and by age twenty had earned her private, commercial, flight instructor, and ground school instructor ratings. She taught future military flyers through the War Training Service program (WTS) at Odessa, Texas, before joining the WAFS. She was number thirteen and had 899 hours.

Helen Richards was the only girl accepted into CPT Advanced Pilot Training in Van Nuys, California, in 1940. When she finished her two-year college course at Pasadena Junior College, she took off for Floating Feather airport near Boise, Idaho, where she earned her commercial and instructor's licenses in 1941 and began teaching. When Nancy Love's telegram reached her, Richards was running the flight school at Floating Feather. When she reported to Wilmington in September 1942, she had just turned twenty-one, had 964 flight hours to her credit, and was the tenth woman admitted.[4]

Betsy Hulet Ferguson and her sister, Sally Hulet Duncan, learned to fly in the 1930s, and took their flight tests and received their pilots' licenses the same day: August 26, 1938. They were known as Coffeyville, Kansas's Flying Sisters. Betsy was the first female Kansan to receive her commercial pilot's license. Next she earned her instructor's rating. In

1941, Sally's husband, Hank, was granted a government-subsidized CPT contract. He hired Betsy and Sally to teach college students to fly through CPT and later the WTS programs at Coffeyville Junior College. With 873 flight hours to her credit, Betsy headed for Wilmington to meet Nancy Love and became WAFS number twenty-three.[5]

Dorothy Scott enrolled in the University of Washington's CPT program in January 1941. She soloed on March 17. After college, she earned her instructor's rating and, in fall 1942, was teaching flying to military cadets at Washington State College in Pullman when she heard about the WAFS. She reported November 20 and was number twenty-five, the last one to join the squadron in 1942. She had 504 hours and got her 200-horsepower rating in a Waco (pronounced Wah-ko) just before she left Washington for Wilmington.[6]

The job of welcoming the WAFS to the 5th Ferrying Group fell to base commander Col. Thomas D. Ferguson. Nancy Love's reputation preceded her. Her background and her standing with the powers-that-be in the Air Transport Command assured her a warm welcome. One of her strengths was her ability to walk into any Operations office in the Ferrying Division and give help wherever and however needed—and with assurance and finesse. Colonel Ferguson was aware of this.[7]

When Nancy Love got to Dallas, she immediately checked out on the AT-6, the advanced trainer built at the North American plant in nearby Grand Prairie. She soloed on January 8, and on January 11 ferried an AT-6 to Romulus, Michigan, where she helped Del Scharr set up the second women's squadron. She also got in some B-24 copilot time at the Ford Willow Run aircraft plant across the road. In February, she moved on to Long Beach, California, to help B.J. Erickson set up the women's squadron there.

Photo 23. Nancy Love was the first woman to fly the P-51.

Courtesy WASP Archive, Texas Woman's University, Denton, TX.

While Nancy was in Long Beach, she became the first woman on this side of the Atlantic to check out in a pursuit aircraft and the first to fly the P-51. It is noted in her logbook as A-36/P-51 with a 1325-horsepower Allison engine. This was an older model than the high-altitude, high-performance P-51 C and D equipped with a Merlin engine that the women pursuit pilots of the Ferrying Division would fly throughout most of 1944. Her instructor, Major Samuel C. Dunlap III, was an old friend of both the Loves. At that time he was the Operations Officer for the 6th Ferrying Group at Long Beach.

Nancy familiarized herself with the technical specifications and early on the morning of February 27, 1943, she and Dunlap met at the end of an isolated runway. He was standing by the A-36. He gave her a cockpit check and waved her off. "I felt the same lonely but wonderful feeling you get on your first solo."[8]

Major Dunlap signed off in her logbook. "Qualified at Long Beach, California."[9] A woman had just flown one of the fastest airplanes ever built.

Since many of the airplanes the women would be flying were built in the Los Angeles basin, Nancy opted to transfer to Long Beach in order to have access to these airplanes.

On March 4, 1943, Delphine Bohn, original WAFS number fifteen and a native of Amarillo, Texas, was transferred from Wilmington to Dallas to replace Nancy in the squadron. Delphine had soloed a Piper Cub in April 1940, at English Field in Amarillo, then gone on to earn her commercial and flight instructor's ratings. She, too, was a CPT flight instructor prior to joining the WAFS.[10]

Helen Richards's progress up the transition ladder at Dallas appears to be the most rapid among the WAFS stationed at Love Field; therefore it is a good gauge for how the women moved through transition that winter and spring. Helen flew her first BT-15 at Love Field on January 8. She delivered her first BT-15 to Enid, Oklahoma, on January 14. Her first overnight in that aircraft was to Greenwood, Mississippi, January 21. Her first flight in an AT-6 was February 15 and she delivered several

BT-13s during that month. Her second AT-6 flight was March 16, and she delivered more BT-13s throughout March and April.

In May, Helen added a checkout in the Stearman PT-17 and soon after that began flying the twin-engine AT-9. Ferrying PT-17s kept her busy most of June and she also added to her AT-6 time. July brought long-distance AT-6 deliveries and also time in the twin-engine AT-11 and the twin-engine UC-78.[11]

Helen proved to be one of the Ferry Command's most reliable pilots.

On May 1, the first six WFTD graduates reported to Love Field and the women's squadron. They were Mary Lou Colbert, Edna Collins, Byrd Granger, Evelyn Greenblatt, Ann Johnson, and Sidney Miller.

Edna Collins had received a call from Jackie Cochran to participate in the training program. Byrd Granger owned her own public relations firm in New York before the war. Evelyn Greenblatt learned to fly in secret because her parents didn't approve. She met fellow pilot Ann Johnson, a doctor's wife, and the two of them became the first female air traffic controllers in Georgia. Both joined the Civil Air Patrol in Atlanta and they entered Class 43-1 together. Sidney Miller made her debut at the Cotillion in Fort Worth in 1934 and attended the University of Texas. She came from a ranching family and her father was mayor of Mineral Wells, Texas, for ten years.[12]

Byrd Granger, 43-1, wrote about her early ferrying experience out of Dallas in her book *On Final Approach.* She describes how the newcomers, in order to learn the proper procedures for delivering aircraft, were placed under the guidance of the WAFS, who served as flight leaders. On May 21, 1943, they were all flown to the Boeing factory in Wichita, Kansas, to "buy" Stearman trainers, which were to be flown to training bases and "sold."[13] That was the terminology used. When you picked up a plane, you "bought" it. When you delivered it successfully you "sold" it.[14]

Granger also describes their ground school education: filing flight plans, checking weather, preflighting the PT-17s, climbing into cockpits, and waiting for the lineman to crank up the engine. "It catches hold with a roar and, with the flight leader heading out first from the parking area, the women pilots follow like ducklings after a waddling mother. On their way from the ramp to the runway, the bright yellow planes make S turns right-left-right-left, slowly, under control, all the time eyeing the plane ahead."[15]

Eleven women from Class 43-2 reported to Dallas in June: Betty Bachman, Ruth Dailey, Frances Dias, Betty Eames, Jane Emerson, Kay Gott, Ross Kary, Avanell Pinkley, Helen Ricketts, Cappy Vail, and Betty Whitlow.[16] These women followed in the footsteps of the WAFS—who served as flight leaders for them – and of the "Guinea Pigs" who had graduated a month earlier.

Chapter 10

2nd Ferrying Group, New Castle AAB, Wilmington, Delaware

Betty Gillies, the first woman to join the squadron, was Nancy Love's choice to run the Wilmington squadron upon her departure January 1, 1943.

Betty Huyler (Gillies) began flying an OX-5 Travelair November 10, 1928, soloed on December 23, and earned her private pilot's license on May 6, 1929. She bought a used DH-60X Moth "to build up my time for my commercial and transport licenses." Because of that newly minted license, she was present when the Ninety-Nines were founded, November 2, 1929, and can be seen in the famous photograph taken that day at Curtiss Field, Valley Stream, Long Island. As one of the ninety-nine charter members of the organization for licensed women pilots, she is part of history. She also served as the organization's president, 1939-1941. She had 1,395 hours and was one of the few to carry a multi-engine rating when she entered the WAFS.[1]

Betty turned thirty-five on January 7, 1943, a few days after she became squadron leader at New Castle. She had been married thirteen years and was the mother of two. She commanded the respect of the women serving under her. Several women who served under Betty in Wilmington say that she was a good listener—a trait Nancy Love wasn't known for—but that she was almost as stingy about handing out compliments as Nancy was.

The first event under Betty's regime was the addition of Helen McGilvery and Sis Bernheim to the squadron. Both passed their check rides and were welcomed the first week in January.

Helen Elizabeth Schmidt was a 1937 graduate of the University of Michigan. She learned to fly in 1940, and went to work teaching future Army pilots to fly in 1942 at the Bloomsburg, Pennsylvania, airport. There she met and married fellow pilot Charles McGilvery. When he went to Wilmington to serve as a ferry pilot, she went with him—hoping to find a job. Helen—nicknamed "Little Mac"—had 940 hours, 650 of them as an instructor.[2]

"Sis"—born Kathryn L. Strouse—was married to Maurice Bernheim, but was widowed by the time she started flying. She soloed on November 14, 1934, in a Taylor Cub and obtained her private pilot's license shortly after that. In 1938, she became a partner in a flying business, which was "going well until the day the Japanese bombed Pearl Harbor." In January 1942, Jackie Cochran sent Sis to Montreal to take the flight test to fly with the ATA in England. "The check pilot acted like a real pompous jerk," Sis recalled. "He made no secret of his belief that women were lousy pilots. Before the day ended, he made sure he had proven his point by flunking all six American women candidates [including another future WAFS, Barbara Poole] who were there to take the test. I returned home very disappointed."[3]

Betty and eight others had been left in Wilmington to form that core squadron: Nancy Batson, Delphine Bohn, Helen Mary Clark, Dorothy Fulton, Teresa James, Esther Manning, Gertrude Meserve, and Esther

Nelson. With the addition of McGilvery and Bernheim and the departure of Aline Rhonie December 31, this put Betty's squadron at eleven.

Nancy Batson, from Birmingham, learned to fly in the CPT program at the University of Alabama in 1940. She earned her instructor's rating but, unable to land a job at home, became an air traffic controller for Pan Am in Miami. When she became bored, she went to Embry-Riddle Aeronautical Institute, where she got a job as a flight instructor. When she read that the ATC was hiring women ferry pilots, she headed for Wilmington to talk to Nancy Love. She had 503 hours and was number twenty.[4]

Helen Mary Clark's two sons, Gerry (12) and Bill (10), thought having a "fighter pilot" for a mother was, in today's vernacular, "cool." By 1942, she had been flying for eight years, owned her own plane—a Monocoupe—and had logged 629 hours. Her husband was draft-exempt—involved in wartime government work—and they had a faithful housekeeper, thus leaving Helen Mary free to serve when her friend Nancy Love called. She was number five.[5]

Dorothy Fulton learned to fly through a revolutionary two-year flying course established and taught at Teaneck (New Jersey) High School. She soloed March 27, 1936, and earned her license August 5, 1937. She went on to get her commercial, multi-engine and instructor's ratings and taught flying and ground school at Bendix Airport, which today is known as Teterboro. She attended New York University, studied aviation, then went into the business. She bought her own airplanes and ran a flying school at several New Jersey airports. When she joined the WAFS as number twenty-two, she had 1,966 hours.[6]

Bostonian Gertrude Meserve started flying in high school in 1938 —"Piper Cubs, forty- and fifty-horsepower." When she received her telegram from Nancy Love in September 1942, Gert had her instructor's rating and was teaching CPT/WTS students at the various colleges around Boston. All she needed was her 200-horsepower rating, which she went after immediately. She got it in a Fairchild 24. She had 1,964 hours when she joined the WAFS. She was number twelve in the final squadron.[7]

Teresa James soloed August 3, 1933, in a Travelair and received Private License #31249 on October 12, 1934. Her instructor was sure he could make a stunt pilot out of her, and he did. She began doing twenty-six tailspins followed by a series of loops in an OX-5. Teresa enrolled in an Advanced Aerobatic and Inverted Flying course to earn her Secondary Instructor's rating in order to teach in the Army Cadet program—the kids bound for combat. That's what she was doing when she got her chance to join the WAFS as number eight.[8] By then Teresa had 2,254 hours.[9]

In 1930, Esther Gebbert of Los Angeles, age twenty-one, began flying lessons with an instructor named Arthur Nelson. Six years later they were married. By 1938, they had established a flying school in Ontario, California. When Art joined the Ferrying Division in 1942, Esther did the same. Esther had 532 hours and became WAFS number seven.[10]

In January, McGilvery and Bernheim began their thirty-day orientation and the others continued to ferry PT-19s and PT-26s out of Hagerstown—when the weather would allow. Delphine Bohn left March 4 for Dallas, reducing the number to ten.

Then Betty reduced the squadron's ferrying personnel one more. She grounded pregnant Esther Manning Rathsfelder. Esther had concealed her condition by wearing her WAFS overcoat, but when she could no longer get the stick back in the tight quarters of the PT cockpit, Betty put her to work in the office. Esther had been a WTS instructor just prior to becoming WAFS number nineteen. Now she became Betty's administrative assistant whose job it was to handle the squadron's paper work.[11]

This became a real plus for Betty because it allowed her to get out and ferry instead of being tied to the office. Having Betty freed up for ferrying duty was critical to the advancement of women in the Ferry Command for a far more important reason than having her deliver 175-horsepower primary trainers. Betty Gillies was the second woman to fly a pursuit aircraft. She took her first flight in a P-47 two weeks after Nancy Love's February 27 flight in the P-51/A-36.[12]

Photo 24. Wilmington. WAFS Betty Huyler Gillies was the first woman to fly the P-47 Thunderbolt.

Courtesy WASP Archive, Texas Woman's University, Denton, TX.

"It was mid-winter when I started my transition, working toward the P-47, and lots of time went by between flights. In between snowstorms and ferrying missions I got in about ten-plus hours in the AT-6. Transition consisted of about four hours and forty minutes of dual landings and takeoffs. The rest, solo practice. While I was getting the required time in the AT-6, I also checked out in the AT-9 (which had two 295 horsepower Lycoming engines)."

In February 1943, Betty received orders to go to the Republic Aircraft factory in Farmingdale, New York, and meet with the control officer about transition into the P-47—the biggest of the pursuit aircraft.[13] The P-47 was parked in a hangar. The control officer told her to climb in.

"Then he raised the airplane off the ground so that I could play with the landing gear equipment," said Betty. "He answered every question I could think of so that, when I got in that airplane to fly it, I knew I could reach and handle the equipment. I'm so short, the question was, could I reach the controls." (Betty was 5 feet 1½ inches tall.)

"It really was no problem fitting myself in the airplanes. I sat on a cushion that, with the parachute, put me up plenty high. And the only real long-legged airplanes were those for which I had the blocks. I could use a cushion behind me quite well in all but the P-38, the P-47 and the P-51. In those cockpits, the gun sights were too close to my face if I used a cushion behind me. The blocks Grumman made up gave my legs the length I needed [to reach the rudder pedals]. Grumman also made me a gadget to turn the fuel valve in the P-38.

"After my first solo flight in a P-47 on March 8, I put in about fourteen hours practicing landings at Newcastle [*sic*] and made a required cross-country with landings at two other airports. It was quite difficult getting those fourteen hours in, as it was not very often that Transition School had a P-47 on the line and operating. Besides which, I was off on ferrying missions a good deal of the time. So it was May 4th before I made my first P-47 delivery.

"Once the women gained experience, and the men saw how capable they were, many of the barriers fell away. Checkouts, for the most part, were the tech orders, a cockpit check and a written Q and A test in the transition school. Also, I participated in continuous instrument school while on base—in classroom and in the Link trainer."

Helen Mary Clark followed Betty into the cockpit of the P-47 and the two of them were ferrying P-47s by mid-May 1943. Teresa James checked out in the Thunderbolt on July 4, 1943, the third woman to fly it.[14]

On May 3, in came six of the Guinea Pigs: Eleanor Boysen, Marjory Gray, Marion Mackey, Jane Straughan, Magda Tacke, and Dorothy Young. All had been sent to Wilmington by their own request as all had family on the eastern seaboard. Though originally that was thought to be a good idea, both Dedie Deaton and Betty Gillies would quickly discover that it wasn't. It created unexpected problems. Others from the first class were assigned to bases with proximity to home, but it was primarily Wilmington where it proved to be a problem. Possibly it was because everything is closer on the East Coast and these were women from New York and Washington D.C.—neither very far from Wilmington by train.

Boysen would resign for personal reasons in June and Young in August. Boysen had learned to fly at the University of Arizona. She bought a Fairchild 24, got her license and was instructing in California when she heard Jackie Cochran was recruiting women pilots.[15]

In losing Young, Betty lost the "little general" of 43-1. Five-foot-one Dottie Young, the smallest of the graduates, had earned that title for her extraordinary squadron leadership in Houston. Her classmates called her "a three-cushion job" because she had to have three cushions under her to see out over the instrument panel. Dottie was married to Paul Young, director of general inspections for the Civil Aeronautical Authority (CAA) in Washington D.C. They lived in Arlington, Virginia. She resigned August 31, and soon after, she and her husband moved to Denver.[16]

Marjorie Gray's first flight was on July 4, 1937. She soloed a year later and went on to earn her commercial, instructor, instrument, and seaplane ratings. She was working at LaGuardia Airport as an air traffic controller trainee when Jackie Cochran contacted her to go to England with the ATA. Marge declined, but accepted the WFTD training offer. She had 350 hours when she joined.[17]

Marion Mackey wanted "above all things to learn to fly." She took her early training at a small field on Long Island near Roosevelt Field. By the time she received a telegram from Jackie Cochran offering the WFTD training, she had accrued the necessary 200 hours. Dottie Young's husband Paul—a flight inspector with the CAA—certified Marion for both her private and commercial license tests. Dottie and Marion were roommates in Houston.[18]

Magda Tacke, born in Budapest, Hungary, first flew in 1932 and earned her private pilot's license in 1938 flying an OX-5. Her list of airplanes flown prior to joining the WFTD is extensive.[19]

Nine members of 43-2 reported mid June: Virginia Alleman, Catherine DeBernard, Emily Hiester, Melvina Maier, Rita Moynahan, Marie Muccie, Helen Stone, Ruth Grimm Trees, and Mary Trotman. Catherine left on a medical release in early July.

Marie Muccie asked to be sent to Wilmington because it was close to her home in New Jersey. But it turned out that the proximity did her no good. Like the rest of the new pilots, Marie found out how consuming the life of a ferry pilot really is.

"I thought, hey, I can see my family because it's right there close to Wilmington. Would you believe I never saw my family! We were too busy. When you're in the Ferry Command you're no sooner on the ground than you get orders to go out the next day. I was flying full time all the time. We flew every single day from sun up to sundown. We had no time to ourselves."[20]

Chapter 11

3rd Ferrying Group, Romulus, Michigan

In 1941, Wayne County Airport, southwest of Detroit, was the fourth busiest airport in the country. With the coming of WWII, the airport's peak 1,800 flights per day made it an excellent base for the Army's rapidly expanding ferrying operations and it was taken over by the U.S. War Department. Wayne County Airport became Romulus Army Airfield—named for the small community it bordered.

The airfield became the home of the 345th Air Transport Command, of which the 3rd Ferrying Group was a part. In January 1943, a squadron of five women ferry pilots arrived to serve with the 3rd FG. From that initial five, a total of 108 women pilots would call Romulus their home base over a twenty-three-month span.[1]

By 1942, the Ford Motor Company's new Willow Run B-24 bomber manufacturing plant—including an aircraft assembly facility and adjacent airfield—was in operation across the road from the Romulus airfield. This was the first aircraft manufacturing plant to use Ford's automotive mass-production techniques: a leading technological innovation of the time. The plant won fame by producing one B-24 every 59 minutes.[2]

The primary mission of the male pilots of the 3rd FG was to ferry the twenty-ton, four-engine B-24 Liberators to their initial designation. In addition, light aircraft were built at other sites in Michigan and Ohio and brought to Romulus to be refueled and inspected, then ferried elsewhere. The women pilots moved these smaller aircraft.[3]

Nancy Love flew in from Dallas on January 13, 1943, to deliver an AT-6 to Romulus. Immediately the Michigan weather closed in, grounding her and all other flying for several days. She was there to help Del Scharr start up the new WAFS squadron. The weather did clear up enough one day and she promptly used the AT-6 to flight check her newest—and final—original WAFS pilot, Lenore McElroy.

The wife of a Romulus ferry pilot, Lenore had not come to Wilmington because she had three teenagers at home and could not leave them. Now, she and her ferry pilot husband, Clarence, could trade off being home with the kids, allowing her to become a ferry pilot as well. McElroy was thirty-five-years old, an experienced flight instructor, and had approximately 3,500 hours.[4] She proved to be a key addition. A gifted leader and well liked, several months later she would become the capable and reliable second in command of the squadron.

McElroy officially joined on January 23, 1943. On January 25, Colonel Tunner received a memo from General Arnold that, from that date, the Ferrying Division would employ only the women who had graduated from the Army's Flying Training facility in Texas. This effectively cut off any opportunity to recruit and increase the number of women ferry pilots until the WFTD graduates began to arrive in the spring.

Adela Riek "Del" Scharr was a teacher and, at thirty-five, was one of the oldest of the WAFS. She was the seventh to join the squadron and the sixth listed in the final roster because Catherine Slocum, who had been number six, opted to return to her family after completing her thirty-day orientation. Del soloed on July 24, 1935. In 1940, she earned her instructor's and her commercial pilot ratings and became the first female flight instructor at Lambert Field in St. Louis, Missouri.[5] She had 1,429 hours when she entered the WAFS.[6]

Next in seniority at Romulus was Barbara Poole. The day three-year-old Barbara saw a WWI Jenny biplane make a forced landing in a New Jersey field, she became enamored of airplanes. Barbara had 1,800 hours when she went to join the WAFS, but she did not have the required 200-horsepower rating. She left Wilmington, got the rating, and returned to become number nine of the original WAFS.[7]

Barbara Donahue learned to fly in 1939, soloing a Stinson 105 early in 1940 at Roosevelt Field on Long Island. She was teaching flying through WTS in Houston when she heard about Nancy Love's program. She had just over the required 500 hours and was number sixteen to be accepted.[8]

Phyllis Burchfield learned to fly in California at Mills Field, now the San Francisco Airport. "At the time, the southeast runway ended in tideland and mud flats. Landings and takeoffs were usually in a cloud of seagulls," she recalled. She soloed an Aeronca Chief on December 4, 1934, with Moreau Flight service at Oakland Municipal Airport. She earned her private pilot's license in 1935, flying a Great Lakes biplane. A week after that, she received her commercial license. While flying out of the Oakland Airport, she met Amelia Earhart. Phyllis was instructing in Memphis in fall 1942 when she heard about the WAFS and went to Wilmington to join. She was number eighteen. She had 1,675 hours.[9]

Katherine (Kay) Rawls (Thompson) was a champion swimmer and diver—an Olympic Silver medalist in three-meter diving for the United States in 1936 in Berlin. Iris Critchell (43-2) a friend and fellow competitor in swimming events in the 1930s, describes her as "an all-around athlete, a highly competitive but most gracious sportswoman. She put one-hundred-twenty-percent effort into everything. With her superb coordination and athletic conditioning, flying came quite naturally to Kay."[10] Kay's husband, Ted Thompson, ran a flight school in Fort Lauderdale, and he taught her to fly. When Ted joined the RAF as a ferry pilot, Kay followed suit and applied to Nancy Love to join the WAFS. She had 675 hours when she came to Wilmington and was number twenty-one to join.[11]

Those were the WAFS sent to Romulus. With the addition of Lenore McElroy soon after they arrived, the squadron was complete—until such time as the WFTD graduates from Houston began to join them.

Thirty-five-year-old Lenore—known as Mac—was the oldest of the WAFS. She began her flying career in 1930 with a ride in the Ford Tri-Motor that took her over Niagara Falls. She was too frightened to look out the window. That just wouldn't do, she decided, and opted to learn to fly. She married fellow pilot/flight instructor Clarence McElroy in 1935. He was her second husband and she was, by then, the mother of three. The family lived in Ypsilanti, Michigan, just down the road from Romulus and both flew for the 3rd Ferrying Group throughout World War II.[12]

The WAFS reporting to Romulus received a chilly reception—from both Michigan's winter weather and the men stationed there. In January 1943, their commanding officer was Colonel Carlyle Nelson. Colonel Ponton D'Arce replaced him in May 1943.

The women were to fly L2-B Taylorcraft and L5 Stinson liaison aircraft; Fairchild PT-23 and PT-26 primary trainers; and AT-6 North American and AT-19 Vultee advanced trainers. All were single-engine airplanes, from the 65-horsepower Taylorcraft to the 600-horsepower AT-6. However, when the weather permitted them to fly at all, the women quickly found themselves relegated to the smallest aircraft. They didn't see the cockpit of an AT-6 until May.

Photo 25. Romulus ferry pilots

Courtesy WASP Archive, Texas Woman's University, Denton, TX
From left, ferry pilots Claire Callaghan, Elizabeth McKinley, Marjorie Ketcham, all Romulus WAFS 43-1, Squadron Commander Adela Scharr, Lieutenant Fairfield, Lenore McElroy WAFS, Vega Johnson and Margaret McCormick 43-1—an Aeronca delivery.

On March 25, the women of the 3rd were blindsided by the Romulus brass responsible for their fate. The following flying restrictions were issued:

- The women of the 3rd would be allowed to fly only light trainer aircraft. This meant no transition into basic or advanced trainers or twin-engine aircraft
- Women would not be given assignments as copilots on bomber ferrying missions
- Women would not be allowed to transition (add to their experience) on any high-powered single-engine or twin-engine aircraft. (Note: Numbers two and three meant women couldn't "build time" or work

on transition in different airplanes as copilots to male pilots other than their instructors.)

- Women would make their deliveries on alternate days from the male pilots. And they were to be—if at all possible—sent in opposite directions.

No mixed flight assignments or crew assignments would be tolerated.

The director of operations for the Ferrying Division penciled the following comment on a copy of the changes: "Mrs. Love objected to this directive."[13]

With little time to recover from that slap in the face, the women of Romulus AND the female ferry pilots at all the other bases were hit with another set of restrictions, these from Ferrying Division headquarters.

On March 29, group commanders received notice that no woman was to be assigned to flying duty while pregnant and that women were not to fly one day before through two days after their menstrual period. That restriction meant a wasteful eight or nine days of non-flying time per WAFS pilot per month. The pregnancy issue directly affected only one woman—Esther Manning Rathsfelder, stationed at Wilmington—and her C.O. Betty Gillies already had grounded her and put her on desk duty. No one questioned the wisdom of that edict. However, long-standing presumptions on the part of men as to the capabilities of women were at the root of the second. The women were both stunned and incensed. They also had some good laughs talking among themselves. How would such a ruling be enforced?

Nancy Love went over Colonel Tunner's head to the chief of staff of the Air Transport Command, General C.R. Smith (who happened to be a personal friend), and complained vigorously about both directives.

In response, in an April 17, 1943, letter, General Smith wrote that limitations had been placed on women pilots without regard for their individual professional qualifications; that their ferrying activities were now restricted only to trainer aircraft; that they were prohibited from flying as copilots with male pilots on ferrying missions; and that they were being denied advancement into higher horsepower single-engine aircraft or any twin-engine aircraft. "All pilots, regardless of sex, [should] be privileged to advance to the extent of their ability in keeping with the progress of aircraft development."[14]

Tunner was not pleased with Mrs. Love's actions. However, as the military historian/author Colonel LaFarge states, "If she had observed channels, the whole matter would have been killed in Ferrying Division Headquarters and would never have reached ATC."[15]

On April 26, Colonel Tunner sent a letter in which he rescinded the Romulus directive of March 25 forbidding the use of women as copilots with male pilots.[16] Still, to keep some barriers between the sexes in such matters, the directive was reworded to say that, normally WAFS cross-country checkouts would be given by other fully qualified WAFS "when and if available." And, "Except during training, no WAFS is to fly as copilot with a male pilot, nor is a WAFS to be assigned a male copilot. – However delivery flights may be considered training flights when they are planned as such."[17]

After that, the women of Romulus—as well as the other bases—were allowed to transition into flying heavier, higher-horsepower, and more complex aircraft. Some women of the 3rd Ferrying Group eventually served as copilots for the giant four-engine Liberators.

Tunner's letter also rescinded the letter of March 29 relative to the WAFS not flying during periods of "physical disability."

In a May 1 endorsement, the Air Surgeon, in response to an ATC inquiry, pointed out that there was no general restriction placed on the women pilots in the Flying Training Command during their menstrual periods. The matter should be handled locally, he said. However, he noted that he felt pregnancy was a disqualifying condition.[18]

Problems solved. Air cleared. Nancy Love had proved her mettle in handling the needs of her women ferry pilots. And Colonel Tunner's displeasure didn't last long. She was his ace in the hole when it came to dealing with the women pilots under his command.

Six members of 43-1 arrived in Romulus on May 1, 1943. They were Gertrude Brown, Claire Callaghan, Vega Johnson, Marjorie Ketcham, Margaret McCormick, and Elizabeth McKinley. The squadron doubled overnight.

Gertrude Brown and her husband, James, had decided to learn to fly in 1935. By the time the opportunity to become part of WFTD came along, "Brownie" had 334 hours. Her instructor at Houston was Helen Duffy. They had known each other in New Jersey and Duffy told her she would be "a lot tougher on you than if you were a stranger."

Claire Callaghan earned her pilot's license in October 1939 and her commercial and instructor's ratings in 1941. She taught flying prior to entering training in Houston. Margaret McCormick was a pilot and a parachutist in aerial exhibitions before WWII. Elizabeth McKinley was a graduate of the University of Maryland, where she learned to fly through the CPT program. She left the WASP program in February 1944, joined the Red Cross, and went overseas. Vega Johnson was from New York City. Fall of 1943, she was sent to Palm Springs and then Long Beach and ended up ferrying out of Palm Springs in 1944.[19]

The new ferry pilots were not happy about their impending return to flying liaison planes and primary trainers and said so. They had been flying the big stuff—AT-6s and the twin-engine AT-17—in Houston. So the transition question affected attitudes of both the WAFS and the Houston girls.

On May 14, Scharr, leading the way for her fellow squadron members, checked out in the AT-6.[20] On June 2, she notched the twin-engine AT-9.[21] Donahue and McElroy were not far behind, and Thompson was working hard on AT-6 transition.[22] Poole had dropped behind at that point and Burchfield was gone—nursing her husband George Fulton, who had been severely injured in a training accident. Scharr and Donahue embarked on a race to be first woman to fly the P-39.

Mid-June, twelve more female ferry pilots arrived—these from Class 43-2: Marion Betzler Schorr, Patricia Chadwick, Lila Chapman, Mary Darling, Patricia Dickerson, Ruth Franckling, Ellen Gery, Margaret Ann Hamilton, Jary Johnson, Margaret Kerr, Florence Lawler, and Paula Loop. The Romulus squadron doubled again and stood at twenty-four.

On June 21, Scharr, Donahue, and McElroy checked out in AT-16s—slightly less horsepower than AT-6s—and ferried three of them to California. They were gone for three days. When they returned, both Del and Barbara Donahue resumed their transition path focused on the P-39, *the* pursuit aircraft flown by the male pilots of the 3rd Ferrying Group.

The P-39 Airacobra was a single-engine (1200-horsepower Allison), single-seat, tricycle-gear, pursuit (fighter) airplane built in Niagara Falls by Bell Aircraft. Maximum speed for the Q model was 376 mph, cruising speed 275 mph. It was known as an unforgiving airplane. It was a killer, or so said a lot of the men who flew it or were in training learning to fly it.

Scharr made an exhaustive study of the tech orders. She talked to the male P-39 pilots about their experiences and what they knew about fatal crashes involving others. She noted the relationship of overheated coolant to engine failure. By June 27, she was ready. She sat in the cockpit of a P-39 and passed the blindfold test by successfully identifying every control, knob, and switch in the cockpit by feel. Major Hennessey gave her the green light. "You take P-39 #701 right after lunch tomorrow," he told her.

On June 28, 1943, she did just that. She took the P-39 up and in doing so became the first woman to fly that single-seat aircraft. She wrote down her time upon landing, 1:05. She had successfully notched her first pursuit aircraft. Barbara Donahue was in the air a few minutes after Del, taking her solo turn in the same aircraft.[23]

When Del delivered her first P-39, she took out her lipstick and wrote "Nancy Love" on the fuselage. Her second delivery, "Betty Gillies." Her third "Cornelia Fort" and so on through the list of the twenty-eight original WAFS in the order they had joined.[24]

Gertrude Brown, writing Jackie Cochran about another matter, added this note: "We are being kept busy, seldom staying at the base more than a day or two. Most of us have checked out on the AT-19 and AT-6, although we continue to fly L-5s and PT-26s."[25]

Chapter 12

6TH Ferrying Group, Long Beach California

When the Western Division of the Air Corps Ferrying Command was organized on August 1, 1941, the Long Beach airport became its home. When the Air Transport Command replaced the Air Corps Ferrying Command in 1942, the Western Division became the 6th Ferrying Group, one of the critical cogs in the worldwide airline the ATC was beginning to build. Pilots from the 6th would ferry airplanes around the globe throughout World War II. Colonel Ralph E. Spake was the C.O. when the women arrived, but Colonel Andrew B. Cannon replaced him in August 1943.[1] So many aircraft factories producing warbirds—North American, Lockheed, Vultee, Northrop, and Douglas among then—were located in the LA Basin, and the 6th grew to be the largest of the ferrying bases with more than 1,700 pilots.

Barbara Jane "B.J." Erickson was named the C.O. of the squadron of five women pilots who took up residence with the 6th Ferrying Group at the Long Beach Municipal Airport, February 15, 1943. Accompanying her to Long Beach were Bernice Batten, Cornelia Fort, Evelyn Sharp, and Barbara Towne.

Erickson came from the University of Washington CPT program. She had completed all four available CPT classes, including advanced instruction. "Girls weren't supposed to get secondary, cross-country and commercial," she said. But B.J. did.

After acquiring all of her ratings, B.J. returned as an instructor for the University of Washington flight school. However, when all flying was moved inland after Pearl Harbor, she was out of a job. She still had a semester of college to complete, so she could not move inland to instruct. The Seattle native took a job at Boeing as part of the swing-shift team doing wing assembly on the B-17 Flying Fortress. When she graduated that spring, she did go inland to Walla Walla to take a flight-instructing job. That's where she was when she received a telegram from Nancy Love asking her to come to Wilmington and apply for the women's ferrying group. Barbara was one of the youngest in the squadron, but Nancy Love quickly recognized that not only could Barbara fly, but she was a born leader. Nancy asked her to lead the women stationed at Long Beach.

In two years' time, B.J. went from helping build B-17s to becoming the third woman to fly the Fortress. But before she got her chance to fly the B-17, however, she had a squadron to build and lead in Long Beach.[2] B.J. served as C.O. at Long Beach the entire time the women's squadron was active. To this day she says, "I was the luckiest girl alive. I was in the right place at the right time."

Cornelia Fort soloed April 27, 1940, and earned her private pilot's license in June and her commercial and instructor's ratings by March 1941. Those skills led her to Hawaii where she taught flying at John Rodgers Airport outside Honolulu. She was in the air with a student the morning of December 7, when the Japanese bombed Pearl Harbor. She and her student made it safely back to the airport. Back on the mainland, Cornelia received a telegram from Nancy Love on September 6, 1942. She became Nancy's second recruit for the WAFS.

Evelyn Sharp grew up in Ord, a small rural Nebraska town. Her father's boarder, Jack Jefford, noticed her interest in his hangar flyin' stories told around the dinner table. He was in arrears with his room and board and offered John flying lessons for Evelyn in exchange for what he owed. She was fifteen when she took her first lesson, February 4, 1935. She soloed thirteen months later and went on to earn all her ratings—capped by her transport license at nineteen. When the opportunity to join the WAFS came along in fall 1942, Evelyn was in Nevada teaching future Navy airmen to fly through the CAA/WTS program. In October, she got her release and headed east to fly for Nancy Love. She was number seventeen and had 2,968 hours.

Twenty-year-old Barbara Towne began taking flying lessons in 1936 and had already soloed before her father discovered what she was doing. Paying for flying lessons was not a problem for Barbara's father, but getting her the right kind of instruction was. He sent her to Ryan School of Aeronautics in San Diego to finish her training. After that, she went to work teaching flying in Ryan's CPT program.

Barbara had recently married an AAF pilot when she read about the group Nancy Love was forming in Delaware. With her husband in the service and not content to sit home for the duration knitting for the troops, she headed east and became number eleven of Nancy's WAFS squadron.[3]

In May 1928, fifteen-year-old Bernice Batten saw a barnstormer flying near her hometown, Pittsburg, Kansas. She and her father were the pilot's first passengers that day. Bernice was hooked. She left Kansas State Teachers College after two years to take flight training and worked part-time to pay for her lessons. She earned her private license on August 22, 1934, her limited commercial on September 19, 1934, and her transport license on March 5, 1936. Bernice and two other WAFS-to-be—Barbara Donahue and Esther Manning—were CAA/WTS instructors in Houston in September 1942 when they got word about Nancy Love and the squadron she was forming. Bernice was number twenty-four to join the squadron.[4]

Nancy Love took a permanent transfer from Dallas to go to Long Beach, where she helped B.J. get that squadron started. While there, she notched two impressive checkouts—the P-51/A-36 (an early version of the P-51) and the twin-engine workhorse of the AAF, the C-47, known in the civilian world as the DC-3 airliner.

Very quickly, the Long Beach WAFS transitioned into the basic trainers —the BT-13s and 15s that the trainees in Houston had recently begun flying. These were built at the nearby Vultee factory. Soon the five WAFS joined the many male pilots ferrying the BTs to training bases located across the southern tier of states. Frequent trips were to Dallas or Enid, Oklahoma, which meant flying the 450-horsepower aircraft across the southern extension of the Rockies and a lot of southwest desert.

But it was out over west Texas—oddly enough, not far from Sweetwater, where the women were just beginning to train—that tragedy struck. One of the young male pilots ferrying a BT in a group that included Cornelia Fort did some aerobatic maneuvers and came too close to her airplane. The fixed gear of his airplane struck her left wing. The tip of that wing was made of wood. The impact peeled six feet off the wing's leading edge.

Cornelia's plane spun, went into an inverted dive, and crashed vertically into the ground. The initial impact had apparently knocked her unconscious or killed her instantly because she had made no attempt to open the hatch, nor did she try to right the airplane.[5] Cornelia, twenty-four, was the first woman pilot to die on active duty in U.S. history.[6]

Nancy Love, Barbara Erickson, and Sam Dunlap flew to Nashville for Cornelia's funeral. The three of them led the processional up the church aisle at Christ Episcopal Church.[7] More than fifty years later, Nancy Batson—a fellow Original—said she didn't think that Nancy Love ever truly got over Cornelia's death.[8] Nancy Love didn't speak at the funeral but she wrote a personal letter to Cornelia's mother. She also wrote to Cornelia's first flight instructor, Aubrey Blackburne.[9]

That spring, Colonel Tunner told Nancy Love that he was going to need her at Ferrying Division Headquarters, now located in Cincinnati. The WAFS operation was a success and the four existing squadrons were about to double in size with the addition, in May, of the first graduates of the WFTD program in Houston. And when the graduates from the second class arrived, the squadrons would double again. Colonel Tunner wanted Love at Ferrying Division Headquarters to deal with matters concerning the increase in female personnel.[10]

Nancy still had aircraft she wanted to check out on. She had specifically made it part of her assignment to personally check out on the more advanced aircraft available to determine the wisdom and safety for advancing some of her more experienced women ferry pilots into these larger and faster aircraft. Beginning May 5, she logged time in C-53s, C-60s, B-25s, and one UC-78. And with her official recheck in the B-25 under her belt, she and Barbara Towne left for Kansas City on June 20, to deliver a B-25. From there, Nancy headed for Cincinnati to take up residence at Ferrying Division Headquarters alongside Colonel Tunner.

Five members of the Guinea Pig class, 43-1, arrived at Long Beach on May 3: Ruth Hellman, Marjorie Kumler, Geri Lamphere, Lovelle Richards, and Betty Tackaberry. Betty had wanted to go to Long Beach because her friend Cornelia was there, but Cornelia was killed before Betty finished training. She opted for Long Beach anyway.

Ruth Hellman had soloed August 5, 1930, and received her license on September 1. She was eighteen. She had married in 1932 and had two daughters when she left to go to Houston. Her husband, Marco, was a major in the AAF. Marjorie Kumler learned to fly in California. She owned her own plane and earned her instrument instructor's license so that she could teach ground school at Pan American Instrument Flying School. High school senior Lovelle Richards took her first flying lesson January 1, 1937. She soloed in a Fairchild 22. After high school she took an accounting job at Consolidated to earn money to fly. Someone showed her an article about Jackie Cochran and the flight training.[11]

Photo 26. Long Beach ferry pilots, July 1943.

Courtesy WASP Archive, Texas Woman's University, Denton, TX.
Top: Virginia Moffatt, Iris Cummings, Ruth Thompson; Middle: Martha Wagenseil, Geri Lamphere, Zelda Lamer; Front: Geraldine Masinter, Barbara Russell, Dorothy Nichols, Lewise Coleman, Carol Fillmore (all 43-2), and Betty Tackaberry, 43-1.

The new girls had barely gotten settled in when, on June 10, ten women from 43-2 reported to Long Beach: Lewise Coleman, Iris Cummings, Carol Fillmore, Zelda Lamer, Geraldine Masinter, Virginia Moffatt, Dorothy

Nichols, Barbara Russell, Ruth Thompson, and Martha Wagenseil. The squadron had quadrupled in size to twenty women ferry pilots.

"On July 26, 1943, we moved into our permanent quarters on base—Barracks T-8 at the west end of all the rows of barracks of the Long Beach 6th Ferrying Group Base," Iris Critchell recalls. "There was a dining hall and an officer's club, which had very good food. At the far end of the base were all utility and maintenance buildings and the armament school.[12]

"Class One and three original WAFS—B.J., Sharpie, and Bernice Batten—were already there. We were housed on the second floor and, as I recall, there were WACS on the base and they lived on the first floor of that barracks. Later when Classes 4 and 6 came to our base, we had to have a second barracks, and this was Barracks T-7.

"Once I was in, I stayed there. My room was on the west side and Ginny Moffatt was next to me on one side. Down at the north end of the hall were Barbara Russell, Lewise Coleman and Carol Fillmore. At the south end, the opposite end of that hall, were Barbara Erickson and Ruth Thompson. Then we had some empty rooms in between. When somebody came out from another base to check out in an airplane locally, there was always an empty room for them to stay in.

"All the showers were in one great big room, and all the johns were in one big great room. Well, we did manage to close off a few of the johns. They were a little bit more private for us. But the showers were all wide open. And that was all one big room across the hall."

Iris recalls that several of the girls were very artistic and decorated their rooms. A sewer pipe vent went up through one girl's room to the roof. It was six or seven inches in diameter—a good-sized pipe. "She took advantage of the possibilities, and painted it red and white—candy striped—all the way up to the ceiling. A lot of the girls painted their rooms all kinds of colors. I didn't paint my room, but I did hang curtains because I was on the sunny side of the barracks. Fortunately, it wasn't real hot, because that's by the ocean, basically cool.

"Downstairs, at one end of the barracks section, was a big empty area —kind of like a large office area. B.J. got it turned into a Ready Room and saw to it that we had comfortable chairs and a couch. There was a counter, a desk behind that counter, and on the wall the bulletin board that kept track of where everybody was every night and whether they were on the way out on orders or returning on orders. Those records were kept by the girls that volunteered to help B.J.—first Bea Medes and then later on Ruth Thompson took care of all the record keeping for the squadron. Probably twenty-five of us were living in Barracks T-8 at that time.

"We were known as WAFS. We were told to say we were WAFS when we were asked what we were. The Commanding Officer arranged that we were to wear the suntans or pinks and greens, and the Air Forces patch, and the Ferry Command wings. And that's what everybody wore, and in the winter, we wore the pinks and greens with the Ferry Command wings. That's all we wore for the largest portion of the year of 1943. We simply joined in with the original WAFS and were called WAFS at the 6th Ferrying Group Base."

Chapter 13

Welcome to the Ferry Command

When the members of Class 43-1 began reporting to the Ferrying Division early in May 1943, General George wrote to Colonel Tunner, briefing him on where things stood with the Training Command and what was expected. He also offered some background on the women pilots of the first graduating class. Their average flying time, upon entering WFTD, was 350 hours. Their training time had, of necessity, been extended to 22½ weeks—primarily because the women, through no fault of their own, were forced to spend their first six weeks flying civilian trainer airplanes with which they already were familiar. During training, they had accumulated an average of 170 hours. Upon graduation, the 23 graduates had an average of 520 hours—on par with the 500 hours minimum entry-level required of the original WAFS.

The 43-1 graduates had been checked out on primary trainers, basic trainers, and the two advanced trainers, the AT-6 and the twin-engine AT-17. They had had "a good course in night flying," and all had soloed at night. In addition, in accordance with Captain Teague's original suggestion, they had been given instrument instruction including Link trainer time. General George concluded his letter by requesting Colonel Tunner to "give the girls a break."[1]

General George, ever the diplomat, was gently letting Tunner know that General Arnold would be watching.

The only official complaint to surface came from Colonel Baker, the 2nd Ferrying Group commander. Baker was a stickler for doing things right. He felt the trainees had not been given sufficient instruction in handling the necessary paperwork involved in ferrying airplanes cross country. He remedied this by setting up a course as part of the new women pilots' on-base ground school to cover the deficiencies.[2] The WAFS had had to learn the same procedures when they entered the service. As the end result, it was suggested that the staff at Avenger help acquaint their upcoming graduates with required paperwork.

When Jane Straughan (43-1) arrived in Wilmington May 3, all she expected to find were her four classmates who had been assigned there.

But there was Betty Gillies. She was so gracious. Later, I told her that I didn't know she and the WAFS were there. She thought that was really funny that we hadn't been told.[3]

We were housed in an officers' barracks that had been turned into quarters for us. Mrs. Anderson—Andy—was living there as a sort of housemother. We had very small individual rooms—side-by-side. Magda Tacke had the one next to me. An iron cot, a built-in place to hang your clothes, and a chair. There wasn't room for anything else.

Betty's room was just four doors up from me, and next to her was her assistant Helen Mary Clark. We all lived on the second floor. The bathroom and showers were down the hall. There were maybe fifteen of us in the barracks at that point. Of course, later, there were more.

The original WAFS treated us well. Later I learned that everybody on base had been told that the women were coming and that they were going to be respected and treated fairly and squarely. That is the way the base was run.

Another example of Colonel Baker's penchant for doing things right —from the beginning.

When we went in the Officers Club, we belonged there. We could eat in the officers' mess and we were treated very, very nicely. Later, I heard it was hard at some of the other bases. It wasn't the fault of the WAFS who were in charge. The military COs were the ones.

Girls at other bases waited and waited to get more additional training to go out on trips. But we didn't. As soon as we had been checked and had gone through the necessary flight tests there at the base—to satisfy them that we could fly—we were sent out with a flight leader.

We took PT-19 Fairchilds and flew in a group with one of the WAFS leading us. We did that once or twice, then we could go out on our own, on orders, and fly alone or along with somebody else.

We got more instrument time—more Link time—but instrument time really didn't mean too much to us. Our restrictions were such that we did not ferry in instrument weather. Now I don't say that some weren't ferrying in instrument weather under emergency conditions, but your day-to-day rule was, you do not fly in weather. You take off one hour after sunrise. You fly. You have an hour for lunch. And you have your gas breaks, depending on the airplane. You don't stop flying until one hour before sunset. So in the summertime it was a long day.

Gen. Barton K. Yount, commander of Army Air Forces Flying Training Command in Fort Worth, asked the Ferrying Division for an evaluation of the WFTD graduates. By late June, Nancy Love had sixty-five WFTD graduates flying for her and she felt she could give him the following informed assessment: [4]

"In general, thorough and well adapted to their duties as ferrying pilots," she wrote on June 26, 1943. "Their attitude and conduct have been generally excellent." She also made some suggestions: additional instruction in the use of the E6B computer with special emphasis on time-speed-distance problems, also temperature, altitude, and airspeed corrections. In view of potential militarization, she suggested instruction in the organization of the Army Air Forces and in military courtesies and customs.

The Ferrying Division had additional recommendations. One was additional practice on crosswind landings in all of the trainers now in use at Sweetwater. The new women ferry pilots had had several ground loops in crosswind landings, particularly when flying in the PT-17 Stearman, which was new to them and which had a much narrower landing gear than the PT-19 they had flown at Houston. Eventually this would bring about the switch from the PT-19 Fairchild the early classes trained on—and known for its wide landing gear—to the PT-17.[5]

The Ferrying Division also recommended more instruction in group flying since deliveries frequently were made in flights of five to nine airplanes. The FD was not asking for regular formation flying, rather instruction in assembling after take-off and holding loose formation to give leadership and help eliminate confusion on first delivery flights. And, lastly, the FD requested more instruction in cross-country flying, including landings at strange fields. The WAFS averaged well over a thousand miles on their ferrying deliveries and the FD believed that the importance of cross-country experience could not be over-emphasized.[6]

Just prior to writing that letter to General Yount, Nancy Love accompanied Colonel Tunner on a visit to Avenger Field. The date was June 16, 1943. Nancy wanted to talk to the future graduates about what to expect in the Ferrying Division.

"You will have the same privileges as the officers," she told them. "You will be entitled to use the Officers Club. It is a very military life. When you are assigned and arrive for duty you will have indoctrination for about one week in order to learn the necessary paperwork you must file when delivering an airplane. You are on duty seven days a week. When you return from a trip you are given a reasonable length of time off to take care of your personal affairs, and your laundry, which has proved to be quite a problem.

"The trips are long, averaging about 1,000 miles. On the trip you are completely responsible for the ship assigned to you, including having a guard and sending an RON [remain over night] to your home base. This RON states where your ship is each night and why, so that your office may know where every ship is every day and every night.

"You will start on PTs and BTs. After you fly a PT 2,000 miles cross country without a radio you will learn a lot about navigation. Then you will go on to AT-6s and AT-17s. It's entirely up to you and your own hard work how fast you progress."[7]

Part IV

Missions Altered

July–November 1943

Between July and November 1943, changes would come quickly within the Ferry Command as well as at USAAF Headquarters. And though the women's program was only a small part of the war's big picture, as a result the entire mission of the women pilots – in particular the women ferry pilots—changed.

What Was Happening in the War?

On the European Front, Allied troops launched their first assault on European, Axis-controlled soil on July 10, 1943. They landed on the beaches of the island of Sicily, which lies in the Mediterranean Sea off the toe of the "boot" that is the Italian peninsula.[1] On September 3, British troops landed on "the toe" of the boot, the Italian mainland. On September 8, Italy surrendered, and on September 9 the Allies landed in force on Italy's west coast and began to make their way first to Salerno, then to Naples and beyond.[2] But by November 15, the Allied forces had reached a stalemate with the Germans. The guns of both sides—along "the Gustav line" between Monte Casino and Monte Camino on the road to Rome—were silent.

In the air war, the Battle of Berlin began November 19 when Britain's Bomber Command launched its winter campaign of nighttime bombing.[3]

In the Pacific War, American troops made a "near-catastrophic amphibious assault" on the Tarawa atoll in the Gilbert Islands on November 20. But by evening, the strike force had established a beachhead.[4]

Chapter 14

Cochran's Power Play

The day the women of Class 43-1 were due to report to their Ferrying Division squadrons, Jackie Cochran put her plan in motion. She sent the following memo to General Arnold. The date was May 3, 1943:[1]

> The more these women pilots are split up and spread around for operational purposes, the more need there is for overall coordination of problems common to them all, such as type of work to which they are assigned, promotions and discipline. You need eyes and ears in whom you have confidence, and they must be feminine. That's the job I would like to do, and which I think I can do well. You and I had this job in mind for myself from 1941 on.

She pointed out that it was General George's proposal that the women ferry pilots should be under the command of the Ferrying Division Sector commander for whom they flew. "He failed to distinguish between operational control and overall supervision of the women and general policies relating to them." This needed a woman's touch, she pointed out, and she was the woman to do it.

> While this is the job I have foreseen for myself from the beginning … I would feel the same way about the need for this kind of overall

> coordination even if you should select someone else. I would rather lose my identity with the women's flying project and see it go well than have all the rank and power you could give me in a set-up that could produce only mediocre results by comparison ...

In her proposal, Arnold must have seen a way to avoid potential problems—dealing with women under his command. This would help make his job easier and allow him to concentrate on the war effort. She would provide a welcome buffer. Give her what she wanted, and he did.

A June 28 memorandum from General Stratemeyer, chief of the Air Staff, to Headquarters AAF, established the office of "Special Assistant for Women Pilots," listed various duties, elaborated on the expected uses of women pilots, and named Jacqueline Cochran to that position. Her official title was Special Assistant and Director of Women Pilots in the Office of the Assistant Chief of Air Staff. At that point Cochran had no say in how the graduates of the WFTD were used once they were in the Ferry Command—that didn't change—but the new memorandum appeared to give her some kind of authority over all the women pilots.[2]

Nancy Love arrived at Ferrying Division headquarters, 309 Vine Street in downtown Cincinnati on June 23, 1943. Colonel Tunner welcomed her—probably with considerable relief—recommended that her title be Executive for WAFS and sent a letter to the Bureau of Public Relations, War Department, stating that Mrs. Love would be permanently stationed in Cincinnati. He also announced that the WAFS were now flying any type of aircraft on which they could check out, including some bombers. "The appointment of Mrs. Love to this Headquarters is in recognition of this past achievement and the future importance of women ferry pilots. Undoubtedly this angle will make interesting material for articles on the WAFS in national publications as well as the fact that women ferry pilots are now flying the heavier type planes."[3]

Both Love's and Cochran's new appointments were announced officially in separate press releases on July 5. Colonel Tunner was promoted to Brigadier General on June 30, 1943, with the official announcement made July 12.

A July 14 memorandum spelled out Mrs. Love's duties: advise Ferrying Division (FD) Headquarters as to the best use of WAFS, plan and coordinate allocation of WAFS of the various Ferrying Groups, and plan and supervise training standards and a progressive air-training program to stimulate interest among WAFS in becoming qualified in more advanced types of aircraft. This, it was stated, would "avoid stagnation on the lighter types"—a key to the new direction of FD thinking.[4]

Nancy's other assignment—obviously designed to relieve Tunner and the male officers on his staff from having to deal with such things—was to formulate "rules and regulations governing conduct, morale, and welfare of WAFS." Nancy wasn't expected to do this alone. She would coordinate with the appropriate FD staff member to execute these duties. As liaison between FD and the Ferrying Group commanders and the WAFS, she would make periodic visits to the four squadrons for inspection purposes. Nancy would work with staff regarding publicity on the WAFS and also plan and coordinate housing for the women pilots of the FD.[5]

The FD/ATC received no explanation as to what Cochran's appointment meant for the WAFS—whether they would fall under her jurisdiction or not.[6]

Cochran was authorized to recruit trainees, make plans and recommendations on militarization, coordinate AAF functions, plans, and projects relative to women pilots, and to make inspection trips and maintain liaison with using agencies.[7] She had much bigger ideas. She wanted to move "her girls" into many different jobs within the Air Forces. And she set out to find those jobs and then convince the men responsible for getting them done that her girls could do the job. These were termed "experimental assignments," and the ATC was told that there would be several such assignments "to ascertain [the women's] adaptability for duties other than ferry pilots."[8]

Cochran termed these jobs *"the flying dishwashing jobs of the AAF"*: jobs the men didn't like doing and considered boring and beneath them.[9] But the women wanted them. It meant they would be flying. That's what the women wanted to do—fly.

The various jobs, other than ferrying, that the women graduates of Avenger Field came to perform in the following months included towing targets for gunnery practice, tracking and searchlight missions, simulated strafing, smoke laying and other chemical missions, radio control flying, basic and instrument instruction, engineering test flying, and administrative and utility flying.[10] Cochran also managed to have groups of women sent to learn to fly the twin-engine B-25 and B-26 bombers as well as the four-engine B-17, all on an experimental basis.

Nancy Love had no desire to do what Cochran had just offered to take upon herself. What Nancy wanted was to be left alone to do her job for General Tunner and the Ferrying Division and fly whenever she could squeeze in a flight. Treating male and female pilots the same and allowing each individual to advance at his/her own speed and capability worked.

"[Nancy Love] was regarded as a great leader because she was a pilot first and an administrator second," Deborah Douglas writes. She was an advocate for her women and was popular with them. "Her personal conduct was always gracious and civil. Most important, she conveyed her pride in the organization without appearing to seek personal recognition."[11]

Love's approach to using women as military pilots worked, but it was not what Cochran wanted. So, as she had done the previous fall when both programs were in their infancy, Cochran set about trying to change the system. She didn't take long to activate her agenda.

She wanted to try some of her women graduates as tow target pilots at the anti-aircraft gunnery training school at Camp Davis in North Carolina. For that, she sought out and won the approval from the newly appointed AAF chief of staff, Major General Barney K. Giles.[12]

On July 17, Nancy Love received a phone call from Colonel Atkinson at ATC. She was to assign twenty-five women pilots to temporary duty (often abbreviated "TDY") under Jackie Cochran. They were to report to Cochran at the Mayflower Hotel in Washington, no later than midnight July 19, 1943. Twenty-three, listed by name, were recent Houston 43-3 graduates who had just arrived at their respective ferrying bases. The other two were to be five-feet-four or taller.[13] Gertrude Brown (43-1) and Lila Chapman (43-2), both from Romulus, were selected. The women had no idea why they were being sent to Washington to report to Cochran.

In Dallas, the six new 43-3 graduates were told to report to WAFS squadron leader Florene Miller. Florene had received a phone call from WAFS Headquarters in Cincinnati. The six newly arrived women were being sent immediately to Washington D.C. They were to tell no one about this. They were going via commercial airliner. It was a special assignment. They were to report to Jacqueline Cochran at the Mayflower Hotel when they arrived—no matter what time of the day or night, Florene added. Their flight out was already booked. That, Florene said, was all she knew.

The six had spent a total of only three days in Dallas with the Ferry Command.

Lois Brooks Hailey recalled, "We're all in Washington, twenty-five of us. We met with Hap Arnold in his office at the Pentagon, and he said, 'We're going to try something new, an experiment in tow target work for the women.' He didn't want us to talk about it, he wanted to keep it quiet, so we'd see if it would work out."[14]

Their mission, Cochran told them, was to show the Army just how much it could expect of women pilots. She went on to explain that they would be flying bigger equipment than they had trained on. One of the aircraft was the Martin B-26 bomber. All chosen were at least five-feet-four-inches tall, bigger women who could reach everything in the cockpit.

They were going to learn how to tow gunnery targets, helping the student gunners at Anti-Aircraft Artillery School learn their job. They would fly many types of missions: tracking, towing, simulating strafing attacks flying at treetop level, and simulating attacks on the beach where they would fly in from the ocean at the top of the waves.

Cochran reminded them that this was an experiment and that they could withdraw if they wished. No one did.

This created a huge headache for the Ferrying Division that had to continue to pay the twenty-five women, even though they were no longer working as ferry pilots. This was the beginning of Cochran's goal to place the WFTD graduates in jobs other than ferrying.

Already, Jackie had notified Major Urban, then the commander at Avenger Field, that she wanted twenty-five graduates from the next class—43-4—for assignment as tow target pilots, not ferrying, and asked him to let her know the girls with the best flying ability. She also asked him to keep the request quiet. She asked Dedie Deaton to send her information on girls the right height (five-feet-four or taller), weight and "deportment."[15]

None of this added to Nancy Love's ease of mind. She was General Tunner's liaison in dealing with the ever-growing women's squadrons within the Ferrying Division. By the end of June, eighty-eight women were assigned as ferry pilots and the number was growing every few weeks. The last thing Nancy wanted was to be permanently stuck behind a desk where she couldn't fly, where her workday would consist merely of tedious administration and paper work and dealing with problems the squadron leaders couldn't handle, but that was now exactly where she was.[16]

Almost immediately, Nancy was faced with a job she didn't want to have to do.

Following a long conversation with Del Scharr, Nancy removed her from her Romulus command. The reason she gave Del—in Nancy's own words, according to Del's memoir—was, "I have reports on you that as the squadron leader you are totally selfish. You are very unpopular. None of the girls like you."[17]

Harsh words. Scharr was stunned. She felt she had been wrongly blamed for things beyond her control. By then, Phyllis Burchfield was missing from the squadron. She had applied for a transfer to Long Beach to be near her husband, who was now recuperating from his training accident. Her request was denied. Phyllis went anyway, and she was released August 4, 1943.[18] The original WAFS were down to twenty-five.

Scharr writes that other requests for transfer, too, had been denied. Most centered on lack of transition at Romulus. The requests had been outside the chain of command.[19] She didn't understand why Nancy Love blamed her for those as well as the general unrest of the squadron. Her writing indicates that she never got a satisfactory answer from Nancy. In spite of Scharr's defense of herself, Love stood firm.

Del's orders were to report back to the 2nd Ferrying Group in Wilmington after a leave that allowed her to go home to St. Louis and spend time with her husband. Nancy Love told her that she had spoken with Colonel Baker and that Del would be put on P-47 transition as soon as she reported to Wilmington.[20] Scharr left Romulus on August 2.

A name change for the women flyers was announced internally on August 5, 1943. One new women-pilot organization—Women Airforce Service Pilots (WASP)—would take the place of the two nominal women-pilot organizations, WAFS and WFTD, and Jacqueline Cochran would head that newly formed organization. It was a *fait accompli*, but the official announcement wasn't made until August 20.[21]

The rationale was sound. Already, women pilots were performing duties other than ferrying, so the designation "ferry pilots" was no longer accurate. But, more telling, Cochran had effectively asserted control over assignments and placement of personnel. Though Love still directed the women in the Ferry Command, she now reported to Cochran in the Pentagon on matters relating to assignment and disposition of personnel.[22]

Betty Gillies told it this way: "We were WAFS until we woke up the morning of August 5 and learned that someone had changed our name while we slept! ALL WAFS WERE NOW WASP."[23]

Chapter 15

Nancy and Betty Fly the Fortress

In July 1943, Nancy Love and Betty Gillies were offered the plum of World War II flying assignments. General Tunner wanted the two of them to ferry a B-17 to England.

Tunner was getting static from male pilots who were wary of ferrying B-17s over the stormy North Atlantic to the United Kingdom and he saw a potential solution. "These flights had become almost routine and there was no reason for complaint. I decided to let a couple of our girls show them just how easy it really was," Tunner wrote after the war.

Though he liked both Nancy and Betty and respected them as pilots, Tunner was not out to make heroines of them. He had used this strategy before. "We were getting a lot of static from pilots who claimed the P-39 to be a flying coffin. ... With no doubts whatsoever, I had a group of girls checked out on P-39s and assigned them to make P-39 deliveries. They had no trouble at all. And I had no more complaints from the men.

"We had scheduled a blitz movement of two hundred B-17s to go over [September 3-4, 1943], and I assigned the two women to one of the planes."[1]

"So many of the B-17 pilots and crews were just kids," Betty Gillies recalled. "They were quite green—quite young and inexperienced—they were being rushed through the training schools. Some had taken their Flying Fortresses off, gone a little way, and gotten cold feet. They turned around, came back, and said the plane was making strange noises or the engines weren't working right. It was nerves. They weren't prepared.

"General Tunner understood that. He and the brass thought that if the girls delivered some B-17s overseas, the young men wouldn't feel that way. The theory had already proved itself with the fighters.

"We were old pilots compared to these kids. They had just graduated with the shortest kind of training ever. They'd get in a big airplane like that—it's complicated and it takes a lot of flights to get familiar with it. Then you add the fact that you're going across the ocean and you're landing in a foreign country."[2]

The massive deliveries of B-17s and their crews to England were flown via the northern route known as Snowball. This route—from Gander, Newfoundland, or Goose Bay, Labrador, to Prestwick, Scotland—was used in 1941 during Lend-Lease to deliver goods and airplanes to Great Britain. Now, in 1943, the U.S. used it to deliver its own airplanes to the AAF squadrons stationed in England.[3]

The Boeing B-17 was characterized as "rugged, stable and easy to fly, it carried the war to the heart of Germany and was universally loved by its crews for bringing them home despite extensive battle damage." The wingspan is 103 feet and the airplane is 74 feet long. Empty, it weighs 32,720 pounds; loaded, 55,500 pounds. Powered by four Wright 1,200-horsepower radial engines, it has a maximum speed of 300 miles per hour at 30,000 feet, and its range is 1,850 miles.[4]

Nancy and Betty spent the month of August learning to fly the mighty Flying Fortress and on August 16, Col. George D. Campbell, Director of Operations and Training/ Ferrying Division Headquarters, signed off on Nancy, Betty, and T/Sgt. Hall as:[5]

> Competent crew to fly B-17 type aircraft on domestic deliveries. They will be permitted to make such deliveries between factories and modification centers, and modification centers and final domestic delivery points on operation orders of your Group. For the Commanding General.

Thus certified, the crew proceeded to ferry and deliver three B-17s to various points in the U.S. Then on September 1, 1943, Nancy and Betty left Cincinnati in B-17F (No. 42-30624) on the first leg of the trip. With them were 1st Lt. R.O. (Pappy) Fraser, navigator; T/Sgt. Stover, radio operator; T/Sgt. Sidney Weintraub, aerial engineer; and T/Sgt. L.S. Hall, assistant aerial engineer. The plane was destined for the U.S. Eighth Air Force in Great Britain. After some delays, they landed in Goose Bay, Labrador, on September 4. Much of their trip from Presque Isle, Maine, had been under instrument conditions.

They never took off from Goose Bay. General C.R. Smith, thinking the two women and their crew were well on their way, sent a wire to England alerting the commander of the ATC European Wing, Brig. Gen. Paul Burrows, that the plane flown by the two women pilots was on its way, and to notify General Arnold.[6]

> Routine delivering flight arriving in your wing will be B-17 number 30624 in few days. For Burrows and Atwood from Smith. Airplane is piloted by WAF crew first pilot Nancy Harkness Love and second pilot Betty Gillies. ... Desire General Arnold be informed.

The telegram was delivered while Burrows was having dinner with his boss, Hap Arnold. Burrows handed the telegram to Arnold who immediately ordered the flight stopped.

The following is Arnold's message:[7]

> Just have seen message from C.R. Smith ... indicating that a B-17 with women crew will leave for England shortly. ... Desire that this trip be cancelled and no women fly transoceanic planes until I have had time to study and approve.

Passengers Nancy Love and Betty Gillies boarded a C-52A in Goose Bay the morning of September 6, 1943. The flight was bound not for Prestwick, Scotland, but back to Presque Isle, Maine. They were going home. Two male pilots took their places and ferried B-17F No. 42-30624 on to Scotland.[8]

"If we had left the day before—when we were scheduled to leave—we'd have been in England before General Arnold ever heard," Betty recalled years later. "But the weather was bad and we couldn't get clearance."[9]

That was the first and only attempt to have WASP pilots ferry aircraft overseas during the war. Stories do circulate that service men saw women pilots delivering airplanes in both the Atlantic and Pacific theaters, but no WASP ever did so—and they will all tell you that, in no uncertain terms. Yes, the women flew to and all across Canada and, though not generally known, two WASP flew to Puerto Rico in December 1944.[10]

Two more original WAFS did check out as Pilot-in-Command on a B-17: Barbara Erickson and Delphine Bohn. It wasn't that the Army *needed* women to ferry B-17s. It didn't. But the Ferrying Division had decided to treat its male and female pilots the same. Gender didn't matter. They were allowing the women pilots to check out in more complex airplanes at their own rate based on individual skill — just like the male pilots.[11]

Jackie Cochran did take notice of the flight and, soon thereafter, sent seventeen graduates from Avenger Field to Lockbourne AAB outside Columbus, Ohio, to learn to fly and check out in the B-17. The women were from Classes 43-5 and 43-6—part of Cochran's new program to move the women graduates into jobs in other areas of the AAF. Thirteen of the seventeen would graduate early in 1944 and go on to fly the B-17 on various gunnery target assignments.

Cochran's move to have WASP assigned to flying jobs outside the Ferrying Division—beginning with Camp Davis in July—was paying off.

Chapter 16

New Mission for the Ferrying Division

In 1942, the fledgling Air Transport Command did not have enough pilots to ferry the high priority aircraft rolling off assembly lines. The Command had to find qualified pilots (men with a minimum 300 hours) and retrain them. Retraining was necessary because ferrying—the delivery of new aircraft— required the ability to fly a different kind of airplane daily rather than concentrating on one. What needed to be moved is what you flew and a pilot needed to be a quick study in the subtle differences of each. Proficiency in cross-country flying was a necessity.

"I set up a program of on-the-job training," General Tunner recalled in his memoir. "Pilots actually performed the mission of the Command at the same time they bettered their flying. Those at the bottom of the ladder would deliver the simplest forms of aircraft, such as artillery-spotting airplanes [liaison aircraft] and primary trainers. As they built up their flying time in these types, they attended ground school and instrument flying school, preparing themselves for the next step up. Gradually, they worked their way from short hops in trainers on clear days to delivering the largest aircraft all over the world."[1]

The twenty-eight original WAFS were given that same training. However, the need for ferry pilots was so dire throughout 1942 and into 1943, FD/ATC also borrowed AAF flight school graduates.

Iris Critchell, Long Beach WASP ferry pilot, explains:[2]

"When we [Class of 43-2] arrived at Long Beach [June 1943], we found quite a few newly commissioned graduates from the regular AAF cadet training schools had been assigned to the Ferrying Division. At that time, the male squadrons were staffed with civilian pilots who had been instructors on the CPT Programs around the country, as well as Reservists called back to active duty from their airline jobs. More than 50 percent of our squadrons—there were seven at Long Beach—were made up of civilian pilots with considerable experience. They were hired to ferry trainers, the main output of the moment.

"Gradually these civilian pilots were checked out and trained by the base Transition Department into whatever aircraft was needed. The former Airline/ AAF Reserve pilots were assigned to C-47s and Lockheed Lodestars and Hudsons, aircraft they had been flying on the airlines."

Critchell points out that these civilian pilots—who were over-age compared to regular cadets—were called Service Pilots. Instead of the ATC wings, they wore wings with an "S" for Service in the middle. As soon as they were commissioned, they were sent on overseas ferry trips as copilots.

"This commissioning of the civilian pilots, over time, allowed the Ferry Command to release most of the regular cadet school graduates back to combat assignments, which is where they wanted to be anyway," Critchell continues. "But early on, the rapidly expanding Ferrying Division/Air Transport Command—operating what amounted to a worldwide airline—was so critical to the war effort, they were given men from the early AAF cadet classes to get started."

Tunner and his staff created the Ferrying Division's Pilot Classification System to rate pilots as to the type of airplanes they were qualified to fly.[3] All ferry pilots—male and female—carried a card stating those classifications.

- Class I—qualified to fly low-powered single-engine airplanes (liaison aircraft, primary, basic, and advanced trainers).
- Class II—qualified to fly twin-engine trainers and utility planes (examples, UC-78, also known as the AT-17, and the AT-9).
- Class III—qualified to fly twin-engine cargo/medium-transport planes (examples C-47, C-60).
- Class III-P--qualified to fly single engine, high-performance pursuits or fighters, the P-51, P-47, P-40, P-39, and P-63.
- Class IV—qualified to fly twin-engine planes in advanced categories, such as attack planes, medium bombers, and heavier transports (examples: the A-20 and B-25).
- Class IV-P—qualified to fly the twin-engine pursuits, P-38 and P-61.
- Class V—qualified to fly the biggest airplanes, four-engine bombers and transports (examples B-17, B-24, B-29, C-54) and able to deliver them overseas.

Tunner stipulated that the women pilots would be classified the same as the men. They could fly any aircraft on which they could qualify.[4] Eventually, Tunner expected to have 500 women pilots in the Ferrying Division. He was concerned that if they were restricted to delivering primary trainers, the FD would be unable to produce any more male ferry pilots. The cross-country experience on the liaisons and the PTs, followed by BTs and ATs, was a necessary step in the transition program. "If any one group of pilots takes that type of experience away from the sum total, we will lose the majority of our training facilities." If the women flew all the PTs built, the men would never make it to BTs and on up. Consequently, the WAFS/WASP would fly everything they were capable of flying.[5]

Unsettling signs of change emerged when the production of trainer aircraft dropped a dramatic 30 percent in spring 1943.[6] This was accompanied by an expansion in the manufacture of pursuit-type aircraft.

September brought another sharp drop in the production of trainers, again accompanied by an increase in production of pursuits.[7]

At that point, the FD/ATC was handed the job of training all of its own pilots. No more AAF cadets on loan. Transitioning male pilots up the ladder to qualify to ferry four-engine bombers or transport overseas was now its primary goal. This specifically applied to male pilots, because women pilots could not be used in foreign operations. Arnold had just made that clear in the wake of the Love/Gillies aborted B-17 flight. Women did not need to reach Class V. And if women monopolized the lower classifications, male pilots would be denied training slots. Up until now, this would not have been a problem due to the high production of trainer aircraft, but that supply was fast dwindling and the need now was for pilots who could ferry big bombers, transport, and pursuit.

Were women ferry pilots, in fact, still needed?

Two realities said they were.

Tunner and his staff had discovered that two critical groups of aircraft could be omitted from the transition sequence that led to mastery of the four-engine aircraft. One was single-engine pursuits; the other, the twin-engine pursuits and attack bombers. The experience ranges of the latter were markedly different from medium and heavy bombers and transports, even though their weight and horsepower were comparable.[8]

As for pursuit, "Although these fast and hard-to-handle planes certainly required more than average experience, the flying of fighters was not in itself of great experience value in working up to the big four-engine planes," Tunner wrote.[9]

FD/ATC needed pursuit ferry pilots, but why waste time and money having male pilots transition into single-engine pursuits, twin-engine pursuits, and attack bombers when flying them added nothing to their training, other than bragging rights—telling people that they were hot pursuit pilots? Why not let the women, who were *not* going overseas and who *were* qualified, ferry those airplanes?

Tunner had his ace-in-the-hole. By August 1943, seven original WAFS already were ferrying pursuits. The Ferrying Division now had 180 WASP on active duty and Tunner had the names and flying records of sixty-four of them—eight more Originals and fifty-six training school graduates—who could be qualified in pursuit, if planes and a facility were available to train them. [10]

FD/ATC now had two missions—plus a challenge: (1) Deliver the rapidly growing number of pursuits to the docks at Newark and in California to be shipped to combat abroad. (2) Train men to fly bombers and transports overseas as quickly and efficiently as possible. The challenge was to make sure that the women pilots could and would progress beyond Class III (twin-engine cargo and transport types) to Class III-P. Pursuit qualification was the added proficiency.[11]

In September 1943, the Ferrying Division announced its intent to train women as pursuit delivery pilots.[12] This brilliantly practical approach worked, but the repercussions down the line would call in to play the sexist attitudes of the military, the establishment, and the country as a whole. Nevertheless, for one brief shining moment, the WASP of the Ferry Command were given the opportunity to fly their country's most glamorous and powerful aircraft. And they made the most of it.

Aside from reflecting the ongoing change in the status of the war, these events in September 1943 were setting the stage for two more disputes between the FD/ATC and Jacqueline Cochran. One was, the course of instruction at Sweetwater no longer suited the needs of the Ferrying Division.[13]

At the end of August 1943, General Tunner was first made aware of an "excessive number of accidents and mishaps among his women pilots, all involving graduates of the Women's Flying Training Detachment."[14] Ground loops were a major concern. (Ground loops are sharp, uncontrollable turns of an aircraft in taxiing, landing, or taking off.)

The WFTD training program had, until then, undergone few changes. Its mission from the beginning was to prepare the women to be capable of ferrying trainer aircraft, unsupervised, under daylight conditions. But the experience level of the entering trainees had dropped from the 200-hour requirement asked of members of 43-1, to 100, to 75 to 35 between December 1942 and March 1943. Consequently, women were being graduated with fewer accumulated hours and, therefore, less experience. More critical—considering the change in the Ferry Command's mission—the new women flight training school graduates sent to the FC needed to be of potential pursuit-pilot caliber, rather than adequate trainer aircraft pilots. In line with that, the FD formally requested that ATC reduce the quota of WASP assigned to it to fifteen per month for the balance of 1943 and throughout 1944. Preferably, these would be the most advanced and capable pilots available.[15]

General Tunner did seek changes in the curriculum at Sweetwater. He asked for closer scrutiny of all pilots to determine "inherent lack of flying ability" and again stipulated 300 accrued hours for assignment to the Ferrying Division. He wanted more emphasis on crosswind landings and takeoffs and taxiing light aircraft in strong and gusty winds, and he wanted a 2,000-mile cross-country trip, landing at several strange airports with a minimum of two nights spent away from base. Each pilot would be responsible for proper servicing of her aircraft and for pre-flights and inspections. Flight plans and flight logs would be thoroughly checked.[16] From then on, comments on the quality of the new women pilots would be closely interwoven with the revised question of their ability to become single-engine pursuit pilots, not AT-6 pilots.[17]

Yet another critical factor—seldom talked about—entered the mix in July 1943. The AAF ordered reductions in aviation training. Between July 1943 and July 1944, "the student cadet census [was cut] from 31,000 to 3,000. [WASP] were part of that cutback in trainees."[18] This order affected the last twelve WASP classes, from 43-7 (who entered training May 23 and graduated November 11, 1943) through 44-10. (See percentages below.)

The AAF's decision to reduce the number of pilot trainees by a more rapid attrition was based on statistics that showed fewer casualties for American combat pilots in Europe than had been anticipated. This same information would bring about the closing of the primary training schools for the men in January 1944. That action resulted in the laying off of civilian flight instructors. Those same flight instructors, faced with the loss both of their jobs and of their draft-exempt status, wanted the WASP's jobs and made their wants widely known when the news of potential WASP militarization hit the floor of Congress.

This reduction in numbers trained/eliminated had no impact on the women already in the Ferry Command. But, as a result, higher percentages of women entering WASP training were washed out of the last twelve classes than of the previous six.

When the WAFS program was launched in September 1942 and WFTD training began two months later, the push was on to supply as many women pilots as possible to release men for combat assignment. But by early in 1943, the more experienced female pilot population had already been accepted for training or rejected for age or some other reason. The majority of the women entering WFTD training from spring 1943 on had the minimum hours needed to qualify. By summer 1943, the need for more pilot trainees had evaporated.

Strict washout rules were *not* in effect when the first eight classes began training — November of '42 to July 5 of '43. Until then, the Army needed as many ferry pilots as it could train. Class 43-4, which had reported in February, had nearly completed its training when the order came through. That class graduated August 7, 1943.

Beginning with Class 43-7, the odds against earning one's wings had increased. The instructors were tougher on their students because they were told to lower the number of graduates.

Percentage of graduates to the number who entered training in the first six classes:[19]

- Class 43-1 — 29 entered and 23 graduates = **79 percent**;
- Class 43-2 — 51 and 43 = **84 percent**;
- Class 43-3 — 53 and 43 = **81 percent**;
- Class 43-4 — 152 and 112 = **73 percent**;
- Class 43-5 — 124 and 85 = **69 percent**;
- Class 43-6 — 123 and 84 = **68 percent;**

Compare:

- Class 43-7 — 98 and 48 = **49 percent**;
- Class 43-8 — 98 and 48 = **49 percent;**
- Class 44-1 — 99 and 49 = **49.5 percent**;
- Class 44-2 — 109 and 49 = **44.9 percent**;
- Class 44-3 — 96 and 57 = **59 percent**;
- Class 44-4 — 104 and 52 = **50 percent**;
- Class 44-5 — 124 and 71= **57 percent**;
- Class 44-6 — 136 and 72 = **53 percent**;
- Class 44-7 — 98 and 59 = **60 percent**;
- Class 44-8 — 114 and 49 = **43 percent**;
- Class 44-9 — 98 and 55 = **56 percent**;
- Class 44-10 — 121 and 68 = **56 percent**.

War News Reveals Need for Long-Range Fighter Aircraft

An October 14, 1943, raid by the US Eighth Air Force on Schweinfurt, deep inside Germany, made crystal clear how critical long-range fighters (pursuit aircraft) were to the success of long-range bombing missions. The Schweinfurt operation's goal was to bomb ball-bearing factories vital to Germany's war industries. The B-17s inflicted heavy damage, but the raid was costly. The bombers were escorted part way by P-47s, but because those fighters possessed limited range, they had to turn back leaving the B-17s on their own, unprotected. Sixty of the 291 bombers failed to return.[20] Protection all the way to the target and back was essential for any future success.

Chapter 17

The Air Inspector

Under contention from the beginning of the women ferry pilot program was the number of hours needed, first to qualify for training, and second to move into the Ferrying Division. Cochran wanted fewer hours to qualify, whereas the FD preferred to stick with its proven numbers. Up until fall 1943, an incoming woman FD pilot needed to be qualified and able to deliver AT-6 type aircraft "without personal supervision or additional training."[1] Beginning that fall, the goal was for the new women pilots to continue to transition up the aircraft ladder with the goal of flying high-performance, single-cockpit pursuit aircraft.

One-hundred-thirty-four WASP eventually rose in the ranks and qualified to ferry pursuit aircraft. The remaining WASP flew any and everything from the Cubstuff [2] liaisons and tiny single-cockpit PQ-8 radio planes to four-engine B-17s. A few flew B-24s and two flew the B-29. By the end of their service, the WASP, as a group, had flown every aircraft in the USAAF's World War II arsenal.

The WAFS entered service with a minimum of 500 hours. The first WFTD class entered training with a minimum of 200, but earned sufficient hours in training to give them 300. Three hundred was the number of hours the Ferrying Division required of its new male pilots.

After the first class, the women's hours required to enter training began to fall off. That women were accepted with one hundred, then seventy-five, and finally thirty-five hours, was against the Ferrying Division's wishes. Cochran bent the numbers, she had General Arnold's permission to do so, and the FD was told to accept them. Numerous young women who did not have the specified hours, but whom Cochran strongly believed in, turned out to be excellent material. Her intuition often proved to be on the mark. Unfortunately, this practice was turned against her when the subject of militarization surfaced in spring 1944.

Class 43-4 was the largest of the classes admitted to the program—152 entered training February 16, and 112 (73 percent) graduated on August 7, 1943. The thirty-five-hour condition for acceptance was in effect when these young women entered training. Though many had more hours than that, the over-all experience level was considerably lower than for the previous classes. Nor had there been time to incorporate the FD's suggestions for changes in the training regimen. Class 5 began training March 26 with 124 recruits and Class 6, on April 25, with 123. The combination of classes more than twice the size of previous ones and admittance of young women with minimum hours brought out flying deficiencies of varying magnitude.

What this meant was that the women graduates coming to the Ferrying Division from the Women's Flying Training Detachment would have less and less time and accumulated experience. At the same time, the FD faced the necessity of moving women capable of pursuit flying forward as quickly as possible.

General Tunner began to receive complaints from ferrying base commanders concerning the flying abilities of some of the graduates. During summer 1943, the new women pilots had been responsible for sixteen aircraft mishaps on training or liaison-type aircraft —mostly ground loops and bent props. The resulting accident rate was excessive and investigation revealed that all but one were attributed to pilot error.[3]

These incidents were considered serious because they spoke not so much to needing more training, but to their inherent ability to fly. The Ferrying Division suspected some of the women were not capable of ferrying—a very demanding kind of flying.[4]

"Beginning in September and continuing until June 1944, comments on the quality of new women pilots are closely interwoven with the question of their ability to become single-engine pursuit pilots."[5]

Complaints, voiced to Jacqueline Cochran by or on behalf of two Dallas WASP from class 43-4 in a September 22, 1943, telegram, compelled her to take action:[6]

> To: Director of Women Pilots, Pentagon Bldg. Washington DC
>
> There appears to be unfair checks and elimination among W-4 girls by Ferry Command this base. Would like to consult you at your earliest convenient [*sic*] some girls ready to resign.
>
> Class W-4

Following the telegram, three more Dallas WASP pilots—arrivals from the newly graduated (September 11) class 43-5—were eliminated. Cochran quickly requested no eliminations until further notice.

A week earlier, on September 14, General Tunner had written to his superiors at ATC regarding shortcomings among the women pilots he had received over the past three months. He sought to eliminate those "who lack an inherent flying ability and those found unsatisfactory."[7] C.R. Smith, deputy commander ATC, sent Tunner's request to the Training Command with the note that ATC concurred.[8]

Referencing Tunner and Smith's correspondence, on September 29, Cochran wrote to the Air Inspector, Headquarters Army Air Forces. "The Training Command considers the accusations ... of such a serious character that they should not go unchallenged." She asked that an Air Inspector from General Arnold's office investigate.[9]

Gen. Robert W. Harper, Assistant Chief of Air Staff Training, wrote to the Air Inspector on October 1, suggesting that the Training Command and the ATC communicate directly with each other to determine training standards and to make recommendations to Headquarters Army Air Forces for the future conduct of the Women's Pilot Training Program.[10] He added, "this office desires to assist ... and will be happy to make available a liaison officer at any time for the purpose."

Cochran wanted to keep her WASP even if the Ferrying Division could not use them. Already she had made her initial moves to have "her girls" sent to other jobs within the AAF. The problem was, no one had bothered to pass this information on to the FD/ATC.

The FD, not realizing that Cochran wanted these women back to assign elsewhere and having no authority to transfer civilians, did what it thought it had to do. It let them go. Clearly confusion reigned. Cochran felt her girls were treated unfairly and the Ferrying Division tried to explain why its actions were warranted.

The Air Inspector Lt. Col. Dudley N. Outcalt conducted investigations at six bases: FD Headquarters in Cincinnati; the training facility at Sweetwater; three of the four ferrying bases—Dallas, Wilmington, and Romulus; and Camp Davis, NC, where fifty women now served as tow target pilots. He gave flight checks to twenty women pilots and interviewed forty-seven. Four of those flight-checked gave "superior check rides." One was a student at Sweetwater and he ranked her performance as "outstanding." Of the other sixteen, only one was unsatisfactory.[11]

He found varying degrees of resentment against the women pilots at the ferrying bases he visited. But "the highest proportions and most serious effects were evident at Love Field in Dallas." All five women pilots who had been eliminated were stationed at Love Field. Outcalt noted some resentment existed at Wilmington, as well as Romulus.

However, of Romulus he wrote, "Women pilots are treated with the greatest consideration and there is evidence of the greatest degree of acceptance of the program and determination to utilize the women pilots."

This is contrary to reports given earlier by the women themselves. However, it should be noted that both the male and the female command at Romulus had changed long before the Air Inspector's visit.

Two types of resentment persisted, Outcalt pointed out. One was an extension of the resentment of women generally found in peacetime around most commercial and municipal airports. The other, which Outcalt found only in the Ferry Command, was the "fear of [male] Service Pilots that they would be displaced from ferrying duties by women pilots and eventually thrown into combat activities." This bitterness would manifest itself again.

The inspector also found that the director of Ground Training at Ferrying Division Headquarters knew nothing about the ground school curriculum at Sweetwater. The inspector asked the personnel in charge at the school to provide the appropriate officer at FD Headquarters in Cincinnati with the content of the school's ground school course.

Communication—lack of it—was a major culprit and needed to be improved.

Already General Tunner had suggested further changes to the instruction at Sweetwater, intended to improve the flying experience of the women trainees: more practice on crosswind landings; more solo and less dual time; and a 2,000-mile cross-country to simulate actual ferrying conditions.

Outcalt concluded that the elimination methods at Dallas were unfair to the women. The check pilots resented them and applied the rules unfairly during check flights. The women eliminated told the Air Inspector that the check pilots "rode the controls" when they were flying, making it difficult for them to properly control the aircraft. And the check pilots often yelled at them.

Supervisory personnel at Wilmington believed that women generally were not qualified to fly airplanes more advanced than primary trainers. This was the opinion of no less than Colonel Baker himself, as documented by statements made by Betty Gillies—even though the two shared a mutual respect. The result? The women pilots at Wilmington had not had adequate opportunity to advance in transition. It should be noted that Wilmington had no proximity to trainer aircraft other than those primary trainers manufactured at the Fairchild plant.

The Air Inspector's recommendation, as far as flying was concerned, was: admonish the commanding officer at Love Field for discrimination against WASP pilots; and remind the commanding officer at Wilmington that women could be used as copilots with male flyers on heavier and more complex aircraft, thus allowing them to move up in transition.

It took an impartial observer, in the person of the Air Inspector, to point out some rather basic facts that should have been obvious to both sides of the controversy. However, contention and lack of trust and reason had gotten out of hand.

The air was cleared on the matter of eliminations. It was most helpful that the Ferrying Division finally was told that Jackie Cochran wished to use her women graduates elsewhere for duties other than ferrying.

But one more battle was still to come before 1943 was history.

Chapter 18

ATC Holds Firm in Tussle for Command

In a year's time—from November 1942 to October 1943—the course of instruction the women trainees undertook had lengthened, from the four months initially prescribed at Houston to more than six months at Avenger Field. The number of flight hours accrued had gone from 115 to 210, and the academic program from 180 hours to 309.[1]

Now women graduates also would be assigned to the Training Command for operational duties and special training projects outside the Ferrying Division.[2] The Ferrying Division's needs had altered and intensified. It wanted only women who could be moved into pursuit flying—and the quicker the better.

The first of November 1943, Jacqueline Cochran, through her position as Special Assistant in the Office of the Assistant Chief of Air Staff, made yet another attempt to gain control over the WASP in the Ferrying Division.

On November 2, 1943, Nancy Love was in Washington at ATC Headquarters. General George had just received a copy of the proposed "AAF

Regulation and WASP Regulation" from Gen. Barney M. Giles, the new (as of July 1943) chief of Air Staff, asking for his comments. Love immediately called General Tunner to tell him of this new move by Cochran and Headquarters AAF. These proposed regulations were intended to replace the Memorandum of June 28 that had given Cochran "vague powers" as the Director of Women Pilots.[3]

The proposed regulations would put the Director of Women Pilots in charge of every phase of the WASP program—recruitment, training, utilization, living conditions, assignment, reassignment, and discharge. Her "lieutenants"—known as Establishment Officers—would directly oversee squadron conduct and morale on the various bases where WASP were stationed, including the ferrying bases. These non-flying individuals would be designated as WASP. AAF Headquarters, not the Commanding Officer of the Ferrying Base, would appoint squadron leaders. Any inadequacy or incapacity in flying would be reported to the Director of Women Pilots, who would initiate an investigation.[4]

General Tunner immediately sent Major James I. Teague to D.C. to work with Mrs. Love on drafting the FD/ATC's rebuttal. The position of Executive for WASP/Ferrying Division was not even mentioned in the proposed regulation and FD/ATC read that as elimination of the position.[5] Nancy Love would be out of a job.

For more than a year Love had watched Cochran's systematic takeover of what had begun as two dedicated programs with different but connected missions. They had been united on General Arnold's order and had become one program with a dual purpose. One was to supply trained personnel for the other, which would grow in strength and usefulness given its increased numbers. But Cochran wanted her own women's air corps answerable to her and for that she needed a large number of women. The previous summer, she had convinced Arnold to make her director of the women's flying program that, in August, became known as the WASP. Tunner and George held firm on Nancy Love's status as the woman in charge of women in the Ferrying Division. Now it appeared that Cochran once again was attempting to sideline the woman she considered to be her competition. Nancy was fed up and so were her commanding officers.

Love and Teague went to work and drafted a response. They wrote that the proposed regulations were in direct violation of established military chains of command. The using agency (in this case the FD/ATC) should determine what training might be needed; non-flying personnel would not be designated WASP; squadron leadership appointments were at the discretion of the base commanding officer; likewise, dress, appearance, and conduct of squadron members would be handled at the discretion of the base commanding officer.[6]

And they added:

> Since it is believed that only confusion and conflict can result if this regulation is adopted, it is the belief of this Command [FD/ATC] that it would be far better not to use any WASP in the Command than to have them operating under the proposed terms.[7]

Very direct words—and the words were Nancy Love's.[8]

The FD/ATC approved the Love-Teague draft and General George sent it to General Giles on November 3. He added that the Air Transport Command "cannot concur in, and strenuously objects to, the attached proposed Army Air Forces regulations." The letter then pointed out that the proposed plan violated the established military chain of command, and "policies established through the Army Air Forces."[9]

General George backed Tunner and Love all the way. "Any system," he said, "which directs administrative control from headquarters AAF directly into the squadron is intolerable. A commanding officer has a mission to perform and he must have full and complete responsibility." George referred directly to the right to handle the women pilots in the ATC on exactly the same basis as the male pilots.[10]

On November 5, the Ferrying Division presented Headquarters, AAF, with its alternative to the proposed regulations. In it, the WASP's job within the Ferrying Division was defined—very much as it already existed and based on the belief that the WASP were civilians and should be handled accordingly.[11]

The outcome? When the official regulation—AAF Reg 40-8—was published December 21, 1943, the positions of the two women whose leadership was paramount to the success of the WASP were better clarified.[12] Nancy Love's position at Ferrying Division Headquarters was secure. The Staff of the Commanding General of the *using* Command or Air Force "will include" a WASP to be designated as the "Staff Director," the provision stated.[13]

Jackie Cochran's main powers continued to be her authority over the distribution of graduates from the Training School. Cochran would have the authority to appoint her own representatives in WASP squadrons *other than those in the Ferry Command.* In fact, most of the pieces in AAF Reg 40-8 that had concerned FD/ATC were eliminated. The one that remained concerned dress—"pending the issuance of uniforms."[14]

Jacqueline Cochran had been working on new WASP uniforms since midsummer. The official proposed uniform of Santiago blue wool gabardine, designed by Bergdorf Goodman in New York, had received not only General Arnold's approval, but also a thumb's up from General George C. Marshall, Army chief of staff. The uniforms would be ready in the spring of 1944.[15]

The original WAFS were not happy about giving up their uniform. After all, they had worn it with pride since October 1942. Nor did General Tunner wish to see them required to change. But it was to be.

And so AAF Regulation 40-8, as revised, went into the books on December 21, 1943.

ATC historian Lieutenant Colonel LaFarge was surprised that Air Staff even bothered to submit Cochran's proposed Regulations to ATC for consideration.[16] In his own history of the Air Transport Command, he was much more blunt:[17]

"Beginning soon after Cochran was named the Director of Women Pilots ... there was a steady push and pull between the Ferrying Division, through ATC, and Miss Cochran over the clash between its [ATC's] sole desire to get on with its mission of delivering aircraft and training pilots, and hers to conduct a large experiment. ... Some of this conflict could have been avoided if Air Staff had been blunter with its Director of Women Pilots in regard to the limitations of a staff officer's command functions."

LaFarge blamed Air Staff for allowing what became a series of contentious situations and demands to happen in the first place. They let ATC "carry the burden of setting [Cochran] straight" instead of doing it themselves. As a result, Cochran was convinced that the ATC was "rebelliously intransigent." FD/ATC's viewpoint was that "Miss Cochran seemed a wild imperialist. Extremely hard feelings developed all round."[18]

As the eventful year 1943 was barreling toward its conclusion, the direction of the war was changing. Hap Arnold already knew his air losses overseas were not as high as had been anticipated. The tide of the conflict had altered and the U.S. and its Allies were now on a winning track.

Those pursuit airplanes that the women pilots were needed to ferry were proving to be vitally important to the battles that were coming and the swift, timely, safe movement of these crucial airplanes to the docks for shipment abroad was now a prime concern of the Ferrying Division.

It was crucial that the women who were capable of flying pursuit be identified, encouraged, and given every opportunity to transition up until they qualified to attend pursuit school. To do that, they had to work up through a series of transitions.

The ferry pilot looking to attend pursuit school had to earn her Class III rating. This meant she was qualified to fly not only single-engine trainers (Class I) but also twin-engine trainers—UC-78, AT-7, AT-10, and AT-11 (Class II)—and Class III twin-engine cargo/personnel carriers like the C-47 or C-60. She needed an accumulated 500 hours and she needed an instrument rating. Then she would be qualified to enter pursuit school. All of this took time and each woman progressed at her own pace and skill level.

Beginning December 1, 1943, aspiring male and female pursuit ferry pilots began to report to Pursuit School in Palm Springs, California.

Part V

Ferrying Can Be Hazardous to Your Health

Ferrying June–December 1943

With the arrival of the graduates of Classes 43-1 and 43-2, the four existing ferrying squadrons quadrupled in size. Adapting to new surroundings and circumstances can be a problem, but for the most part all the women adjusted—the old-timer WAFS and the new-kids-on-the-block, the pioneering Guinea Pigs, to be followed a month later by the talent-rich women of 43-2.

The training school graduates settled into the ferrying routine, flying whatever trainer aircraft they were called upon to fly depending on where they were stationed. Whether this nomadic life they inherited is what the WASP ferry pilots had anticipated or not is a moot point. That is what they got. Many of their encounters en route around the country were tame, maybe even boring, but some were not. Some were downright terrifying while they were happening, but were served up later to laughter that was a mixture of envy and disbelief among others.

In Chapter 19, ten of those training school graduates of Classes 43-1, 2, 4, 5, and 6 tell some of their adventures. The ten:

- Represent all four ferrying squadrons;
- Eight of the ten remained with the program for the duration. The two who left went home to care for ailing parents;
- Describe ferrying at and from their particular base;

... and most important of all ...

- All left oral histories from which we are able to glean these truly amazing stories of personal fortitude, humor, and flying skill.

Chapter 19

Ten WASP Ferry Pilots Tell All!

These ten women's narratives are told as they appear in their oral histories on file with the WASP Archives at Texas Woman's University or, in Iris Critchell's case, from her memoir-in-progress and conversations with the author.

Iris Cummings Critchell, 43-2—Long Beach[1]

The Ferry Command set certain parameters, rules for ferrying, and the pilots respected and complied with these rules. But pilots were free to make all judgments within the rules. Ferry Command leaders had decided that, at a certain point in a pilot's training and experience, it was time to tell them, "the stick is yours, the ball is yours, you are responsible for that airplane. Now you figure out what you need to do and take it to its destination."

Photo 27. Iris Cummings, Class 43-2, Long Beach ferry pilot.

Courtesy WASP Archive, Texas Woman's University, Denton, TX.

In the beginning, our ferrying assignments at Long Beach were to deliver BT-13s and BT-15s from the Vultee factory in Downey, California, to training fields, most often the Basic Training School in Enid, Oklahoma. I left on my first trip to Enid on July 5, 1943.

Recon vehicles would take a group of male ferry pilots and a few women ferry pilots over to Downey to pick up our BTs. Once you got your airplane, you were on your own. It was up to each of us to get off and to get as far along the route as we could during the day. We were restricted to being on the ground half an hour before sunset and being off the ground in the morning within a half-hour after sunrise.

We had been introduced to all the rules of ferrying in our introduction to the base. This included the procedures about weather delays.

At each stop, the first thing we did was take care of the airplane. This meant issuing refueling orders. Next, you proceeded to the Operations Office and signed in your airplane with the operations officer. Then you found the weather station with Teletype information on current weather and weather maps and usually a weather officer to talk with about the weather ahead.

When this was the last flight for the day for your airplane, you filled out a Western Union telegram form in duplicate, addressed it to Cincinnati Ferrying Division Headquarters stating your airplane serial number, your last name, the airport where you were RONing, and the code for the reason you were stopping—for maintenance or weather delay or simply nightfall, which they would know from the timestamp on your telegram. You received the second stamped copy showing you had filed it. You put that in your briefcase with your other paperwork because it needed to be returned to Long Beach.

By 10 PM every evening Cincinnati Headquarters knew the location of every one of its aircraft and their status.

On that first delivery flight, my logbook shows that I made it from Downey to Tucson in 3 hours and 52 minutes and RONed there. The next day I made it from Tucson to El Paso to refuel and from El Paso to Abilene to RON. Early on the next day, I left Abilene and delivered the airplane to Enid.

When we delivered, we unloaded ourselves and our gear from the airplane and proceeded to the Operations Office to get a signed receipt for the airplane—such and such a serial number—that you brought to them. You gave them a copy of your orders and sent the delivery telegram. With that done, we were directed to a transportation officer who arranged to get us from that base to the nearest public transportation or military aircraft transportation on which we could get back as fast as possible to our base.

The BTs did not carry very high priority and so we had to make our way from Enid via bus to the nearest railroad station, which might be Oklahoma City, and then ride the railroad to perhaps Abilene or Fort Worth to catch an airliner that would get us back to Long Beach. It was not a direct route and took quite a lot of time.

***

Jane Straughan, Wilmington —43-1[2]

When we were first ferrying, we took a lot of the open cockpit PT-19s from the Fairchild factory in Hagerstown, Maryland, down to training fields in Texas. Magda Tacke and I took two to Laredo. We wore khakis to fly in—men's khaki pants and tops—and we had our insignias and our wings. We couldn't carry much in the way of extra clothes. We got our B-4 bags later.

The weather wasn't good so this trip took a good long while. By the time we delivered, neither of us had any more clean shirts left. Magda was returning to base, but I had orders to go to Stuttgart, Arkansas, pick up a trainer, and deliver it to Fort Sumter, South Carolina. Wouldn't you know, we caught a coal-fired train that had windows that wouldn't close and ended up covered in soot.

Photo 28. Jane Straughan, Class 43-1, Wilmington ferry pilot.

Courtesy WASP Archive, Texas Woman's University, Denton, TX.

When I got off at Stuttgart, there was nothing there! No hotels, no motels. I tried to call the base. I called and I called. I thought maybe it was a small training field where everybody had gone home for the night and nobody was there to answer. So I went in this bar. I was filthy dirty and carrying my parachute. I announced, "I have to get a room for the night."

"Why aren't you going to the base?" they asked.

I told them I had called and called the base.

They had a room upstairs that I could have. It was full of cockroaches. It was awful. I didn't sleep.

The next morning I went out bright and early to discover this beautiful airfield. The CO did not believe that I called. The switchboard was supposed to be in operation all night long. He said, "You couldn't have called. Somebody would have come and picked you up."

"I did call," I said. "I sure didn't choose to stay where I did."

"I think you ought to go over to the nurse's quarters. Get them to take you over to the PX because you are wearing the dirtiest shirt and pants I have ever seen in my life!"

They were! It was ninety-degrees outside and I had worn those clothes for about four days. I got a clean shirt, but the PX didn't have slacks. Of course, the CO got to tell the whole tale on me at the mess hall at lunch.

***

Ruth Dailey Helm, 43-2 —Dallas[3]

The men of the Air Transport Command—the ones we were directly under, the ferry pilots at Dallas—accepted us. They treated us like their sisters. They were not military; they were airline pilots and older men who had instructed on many different aircraft.

Some joked that ATC stood for the "Army of Terrified Civilians." They were civilians, not military. We had very little military protocol at Dallas. There was a war on and everybody was trying to do what they could do. I was doing what I wanted to do and getting paid for it—and serving my country at the same time.

The summer of '43, we flew the PT-17s—all the way from Wichita to the West Coast. Then we picked up Beechcrafts at Wichita and took them to places down in Texas. Later, we picked up AT-6s at Grand Prairie and took them wherever they needed to go.

Jary Johnson McKay, 43-2 — Romulus[4]

We usually flew in groups. Our orders were mimeographed. There was a list of pilots, the airplane number each was assigned to, and the names of the pilots—the men by their rank and name and the WASP by their names, all there together. We were civilians and they were lieutenants. I'm surprised they didn't have separate orders for us.

One time they sent us to Newark to take these Stinsons just across the water to Long Island and I did a very dumb thing. It was very bad visibility and I was just going by what I was seeing. I went across the water, I saw land, and thought it was Long Island. I did not check my compass and got totally lost. As a wartime safety measure, they had taken the names off the water towers, so I couldn't identify any towns. I had no idea where I was. I was about to run out of gas and it was getting dark.

I had to land. So I picked out a farmer's field. I was pretty panicky. I didn't drag the field first to see if it was smooth. There happened to be a great big ditch across the middle of it. And so I went off like a ski jump, off the edge of the ditch, and landed on the nose of the airplane and went upside down. I wasn't injured but the plane was. I was in Freehold, New Jersey, not Long Island.

Photo 29. Jary Johnson, Class 43-2, Romulus ferry pilot.

Courtesy WASP Archive, Texas Woman's University, Denton, TX.

***

Marion Schorr Betzler Brown (43-2) Romulus[5]

I ferried a lot of AT-16s from Montreal down to Newark, where they were put on shipboard. —British made, but like the American AT-6. — We flew solo and navigated by pilotage. [Visual ground reference and map.[6]] You gotta draw those maps and know where you are because you don't have radios. You don't have any communication, no one to help you, no radar.

We were coming down from Montreal. There's two lieutenants and me. I had a picture view of the Hudson River. Well, the ceiling started to get lower. I decided to go up, because then I could see what they were doing. When they dove down real low, I knew they were going to fly right down the Hudson River.

I wasn't familiar with that area. I was always afraid that there might be wires across the river, big, tall wires. Well, they went merrily on down below the overcast and down the river. I'm upstairs. I had my course figured out and my time and finally I thought, it's about time to let down through the clouds. I started down.

Suddenly I saw this huge face in front of me. It scared the daylights out of me. All I could think was—it was God! This big, white face. My instinct was to peel off to the right. I did. I turned right, real sharp, and climbed. I climbed back up on top. I'm thinking, "where am I and what was that?" On our charts, the Statue of Liberty is a little bitty dot no bigger than a pinhead. It says "Statue of Liberty" in small print. That's where I had to be!

So I drew another course and heading and tried to let down from there. Sure enough, that's what it was. I flew south a little farther. But I was that close. Then I did my let down and landed.

Afterwards, I couldn't wait to look at a picture of it. The torch is in her right hand. Instinctively I had "peeled off" to the right. If I had turned left, for sure, I would have hit the torch. If I'd gone into the Statue of Liberty, I think she'd have won.

Photo 30. Romulus ferry pilots: Mary Ann Beard, 43-5, Janet Zuchowski 43-4, and Marion Schorr, 43-2.

Courtesy WASP Archive, Texas Woman's University, Denton, TX.

***

Few members of Class 43-3 became ferry pilots. The graduates from Class 43-4 joined their ferrying squadrons in mid-August 1943, and the women from Class 43-5 arrived in mid-September. Consequently, the following stories take place in the fall of 1943 and winter of 1943-44.

Dorothy Kocher Olsen, 43-4—Long Beach[7]

From beginning to end I was so lucky. I've never forgotten one ferrying trip, November 18, 1943. Iris Cummings (43-2) and I took a C-47 from

Long Beach to the modification center in Tulsa. I was Iris's copilot. There, the operations officer asked us if we would like to deliver an A-24 to Newark. I didn't know what an A-24 was. Thank God for Iris. If I'd been with a weak sister, I never would have had this opportunity. My daughter and I call these women that drive in the passing lane below the speed limit weak sisters. Iris is not a weak sister.

He showed us this A-24, gave us a cockpit check, and said, "Okay, take it around the field." No instruction.

Well, bless her heart, Iris climbed in and I climbed in and we checked out. We took it to Newark. Made five stops en route. Just imagine the leg up that gave us. We were over that hurdle. But even better, after that I got to ferry SBDs [the Naval version of the A-24] from the Naval base off San Diego. All this led to instrument training at St. Joe, Missouri, and from there, the next step was fighter school.

***

Gayle Bevis Ewing Reed, 43-5—Dallas[8]

I asked for Dallas, I got Dallas. I was a ferry pilot, which I wanted to be.

We were assigned barracks right next to the men's. An iron cot with an Army blanket and sheetrock for walls. Where they put studding in between, they made little shelves you could put things on. It was awful, but I was twenty-five. Anything horizontal was easy to sleep on. We ate in the mess hall and used the field just like the men.

We did the best we could to look military. We faked out the Army uniforms by buying officer pinks [trousers] and a matching shirt. We did not, however, wear them in the officers' club at night. We changed into a skirt. No one could wear slacks in the officers' club.

We flew out of Wichita, which was a great aviation-manufacturing base. They made Stearman and Cessna and Beech and Boeing—they all were there. They flew us up to Wichita in a bucket seat airplane where we picked up whatever we were flying.

The Stearman PT-17 is a bi-wing primary trainer that cruises at 90-95. I usually full-throttled it. I would take them from Wichita back to Dallas, and on out to California. My best time on that trip was six days. Open cockpit, goggles and helmet! You have wide-open weather conditions to fly there and there's nothing more exciting than being weathered in in Pecos, Texas.

I flew a lot of Stearmans. Later on, it was kind of fun when we were picking up Stearmans because they were out of the same hangar as the B-29s. I'd go in to Operations and say, "I'm here for a B-29."

***

Sylvia Dahmes Clayton, 43-5—Romulus[9]

We picked up PT-26s at the factory in Fort Erie, Ontario—right across from Buffalo. Five of us. The weather was bad. Even with snowstorms, we had to go out to the factory everyday. The bus took us out and we sat there until 2 o'clock, which was the last time we could get two weather reports that were good and still take off and get to the next town. We played cards, we studied, we wrote letters, we read, and we played some poker.

Photo 31. Sylvia Dahmes, Class 43-5, Romulus ferry pilot.

Courtesy WASP Archive, Texas Woman's University, Denton, TX.

The PT-26 was the same as our PT-19, but had a canopy. These planes were going to Calgary and it was wintertime. Finally we took off, got into more snow over southern Canada, and landed in Windsor near Detroit. From there we went on to Winnipeg. We flew along the railroad track, so low that I swear we had to climb when the train came along.

We flew into our alternate airport—Medicine Hat—that was a training base for British cadets. They didn't know what to do with us. The British cadets were not allowed to speak to women when they were in training. We sat in the Commanding Officer's office until they served the cadets their dinner—several hundred of them. They cleaned out the dining room and then took us in and fed us.

Well, it was still snowing, so what were they going to do with us for the night? They put us in the emergency ward in the hospital.

The first thing their people on the ground said to us when we landed was, "we quit flying hours ago."

***

Inez Woodward Woods, 43-4 —Romulus[10]

I had orders to pick up an airplane at Wright Field in Dayton and take it to Cherry Point, North Carolina. It was a PQ-14 that belonged to the Navy.[11] It was a one-place airplane [single cockpit]. They looked through the records and found that I had flown Culver Cadets commercially. Well, the PQs were all Culvers. So they decided that either I was expendable or that I had the best chance of making it.

Photo 32. Inez Woodward, Class 43-4, Romulus ferry pilot.

Courtesy WASP Archive, Texas Woman's University, Denton, TX.

I tracked down Wright Field and ran into one of my few cases of macho.

I signed for the plane and asked this officer for some input. He said he'd flown it.

"What's its take off speed?" I asked.

"Well ... you've got a five-thousand foot runway, go find out," he said.

I said, "Well, do you bring it in with power or does it glide in?"

"Well ... take it upstairs and find out what it does."

I could get no information. Did he have tech orders in the plane?

"No, I don't know where the tech orders are. What are you worried about? It's a good plane. I've snap-rolled it."

You cannot snap roll that airplane. They're placarded against it.[12] They're plywood planes. But I signed for it, went out, got in, ran it up, and took off.

Not too far out, over the Tennessee hills, I watched my oil pressure go down. I landed in this farmer's field.

"What was the matter little lady?" he asked.

I told him I was losing oil. I undid the cotter pins and looked inside. It was a mess of oil. He didn't have any, but he took me down the road to the filling station and I bought oil, came back and poured it in.

Then I walked all over the field. Finally, satisfied I could get out, I took the PQ back in the corner, held the brakes, revved it wide open, and took off. There were trees at the end of the field.

I was soaking wet when I cleared them.

The nearest Army Field was Lebanon, Tennessee. I landed there. Maneuvers were on and I was in the middle of them. The Red and Blue armies were learning about combat.[13]

I didn't want to take off with a defective airplane. I was leaking oil and it was a plywood plane covered with dope.[14] It was a flying coffin. As far as I was concerned, it was a torch! So I decided to spend the night in Lebanon.

There were no hotel rooms and it was getting along towards dark. "Well, I'll just sleep in the Ready Room," I told the officer-in-charge.

"Oh, you can't do that. There's too many men around here. You'll never make it through the night," was the answer. He paused. I waited.

"Excuse me a minute," he said. A few minutes later he came back to me. "I have a place for you to sleep, but I'm going to take you to dinner first."

We had dinner and then he said, "Now, you go use the restroom here and do whatever you're going to do because you're not going to have a restroom until tomorrow morning when I pick you up."

"All right," I said, and did what he said. I had no luggage—a toothbrush was about all—because I was flying a one-place airplane.

So I went with him to this house and we opened the door. There wasn't a soul in sight. We went up on the landing to a winding staircase. There was a door tucked away. He pushed it open. "This is your room. Lock the door. Do not come out until I come and knock on your door tomorrow and give you my name."

I said, "All right."

It was like an oversized closet, but it had a clean bed in it. I was tired after that ordeal. I lay down and went to sleep. But all night long I heard men coming and going up and down the stairs. A couple of drunks even bounced off my door, apparently trying to make it around that corner in the hall. Slowly it dawned on my thick skull where I was. —But I was there, so what about it!

The next morning, the officer knocked at my door, identified himself, and took me to breakfast. I called Berry Field in Nashville, an ATC base about thirty miles away, and asked them to let me have a straight-in approach. When I got there, they ripped out the gas lines as well as the oil lines and rebuilt them.

Oh yes, and they entertained me royally while I was at Berry Field.

Alice Jean "AJ" May Starr, 43-4—Romulus[15]

We—flight leader Margaret Ann Hamilton (43-2), and Janet Zuchowski, Virginia "Tex" Clair, and me, Alice-Jean (AJ) May, all 43-4 and all of the 3rd Ferrying Group, Romulus AAB, MI—were ferrying L-5 Stinsons, made in Wayne, Michigan, to someplace in North Carolina.

One stop was in Lebanon, Tennessee, as routed by Operations at Romulus. As we approached, we could see the grass-covered runway cut through the forest. We all four landed and taxied up to a lone Jeep with a couple of GIs with a radio. They gave us a ride to town.

The town square was very crowded with GIs and many townspeople. Our hotel rooms were not ready, so we decided to sightsee a bit. We had a sidewalk photo taken of the four of us and almost immediately were accosted by two very tall MPs. They questioned us—we all looked crummy at this point. We told them we were waiting for our room and then we followed them to an office on the square. Perhaps they were giving us a place to wait in peace and comfort?

Photo 33. Four Romulus ferry pilots get caught in War Games cross fire

Courtesy WASP Archive, Texas Woman's University, Denton, TX.

Virginia "Tex" Clair, 43-4, Margaret Ann Hamilton, 43-2, Janet Zuchowski and Alice Jean "AJ" May, both 43-4, made an infamous stop in Lebanon, TN.

No, it was the sheriff's office. So we told our story: who we were, what we were doing, etc. We were ushered to the side of room. People kept coming in—many townspeople mixed with GIs. Evidently, they were witnesses against us.

Apparently the sheriff was also mayor and the chief of police. At one point, Margaret Ann attempted once again to step forward and tell them who and what we were and what we were doing there. She was physically shoved back and her purse fell off her shoulder.

We insisted they call our C.O., Colonel D'Arce, back in Romulus. They couldn't reach the Colonel—he was out for the evening. Finally, FINALLY, the call came through.

Byrd Granger, 43-1, wrote in *On Final Approach,* "Colonel D'Arce's voice nearly melted the wires as the Romulus C.O. read the riot act."[16]

The mayor told us to go back to the hotel and no place else, eat dinner in the hotel, go to our rooms, stay there, and leave first thing in the morning. The rooms in the hotel were awful—holes punched through the walls. We were bothered all night by pounding on our door.

"Tex," wearing her boots as usual, was outraged and ready to "shoot the town up." She usually carried a gun and we had to try to settle her down. We were very glad to get out of that place.

As it turned out, the Army was in the middle of maneuvers—war games. We had flown and stumbled into the midst of that and had been captured by the RED ARMY—in Lebanon, Tennessee, USA!

We later heard they were sent overseas. We were all sure that they were well prepared.

Back at Romulus, the JAG [Judge Advocate General] questioned us individually. The result? The officer in charge of this war game was told to write an apology.

***

The women ferry pilots who had joined the Ferry Command between May and September 1943 had, by December 1943–January 1944, evolved into capable, trustworthy deliverers of small aircraft all over the country. But much bigger challenges lay ahead.

Part VI

1944: WASP Fly "the Game Changer"

December 1943–July 1944

As eight WASP and thirty-five male ferry pilots[1] arrived in Palm Springs for the opening of Pursuit School December 1, 1943, the Big Three—Franklin Roosevelt, Winston Churchill, and Josef Stalin—had a fateful rendezvous in Tehran, Iran (November 28–December 4, 1943).[2] A night raid on Berlin by the RAF, December 3, killed more airmen (230) than Berliners on the ground (141).[3]

What would most affect those future pursuit pilots was the December 11 news that the British were producing "drop-tanks" made out of corrugated paper to be fitted on the P-51 fighters the men and women soon would be ferrying to the docks at Newark for shipment to the war zone. These tanks—attached when the aircraft got to England—would give the fighters enough fuel to add 600 miles to their range. This made it possible for the Mustangs to cover the B-17s and B-24s all the way to Berlin, take on German fighters, and have sufficient fuel to cover the bombers on the return to their bases in England.[4]

The P-51s sent to Italy had a similar mission. The men stationed with the Fifteenth Air Force at Foggia would fly those Mustangs equipped with drop tanks as an escort for the heavy bombers on the crucial raids on the Ploesti oil refineries in Romania and back[5]—depriving the Nazis of the fuel they needed to wage war and, ultimately, bringing Germany to its knees. The P-51 proved to be "a game changer" in the European war. Author Barrett Tillman writes: "As a weapon, the Mustang was tops. ... With its speed, maneuverability, and range, it could take the fight to the enemy and engage him on better than even terms. At the same time, Mustangs protected American bombers from deadly Luftwaffe interceptors over any target."[6]

This was the overriding reason the work of the WASP of the Ferry Command would prove to be so important. Ultimately, they became a crucial cog in the overall—and vast—machinery of World War II.

Chapter 20

Pursuit School

The Ferry Command, a child born of World War II, was forever an organization in transition—evolving as its mission changed.

As a WASP ferry pilot in an early class—and one knowledgeable in the field of physics and aeronautics—Iris Critchell had a first-hand view of the growth, adaptation, and change that took place during those war years and she understood what was unfolding before her eyes. She was an active participant in what General Tunner and the ATC established. "When the war came along, the Ferrying Division met needs that arose in interesting ways," Iris says, and she talks about them.[1]

Four innovations were needed. *One* was Pursuit School, taught first at Palm Springs and later at Brownsville, Texas. *Second* was an instrument-training course for ferry pilots. St. Joseph, Missouri, was chosen as the location for the school. *Third* was to establish centers where airplanes could be modified according to the environmental conditions where they were being sent. Several were set up. *Fourth* was a system of getting ferry pilots back to base after they delivered their aircraft.

Instrument training: Until winter of 1943, ferry pilots did their instrument training on base, wedged in between ferrying assignments. It was a haphazard method, time consuming and often frustrating for both student and instructor. The instrument card they earned—what was called the military white card—was a requirement for pursuit school. The idea of a centralized school was that, by sending several ferry pilots at a time to the dedicated instrument school at St. Joseph, those pilots received concentrated instruction rather than the strung-out training available on base.

As for the modification centers, an aircraft going to North Africa and a desert environment required different outfitting and finishing than one being sent to the jungles of the South Pacific or across the North Atlantic to fly bombing raids out of England over Germany. Rather than perform these different finishings at the factory, a complicated task given the variety of modifications needed, ATC established centers around the country to finish the aircraft for their specific destinations and assignments. Ferry pilots flew these aircraft to and from the mod centers.

How do you get ferry pilots back to base so they can deliver another aircraft?

In the early days of ferrying, when both men and women pilots were delivering trainer aircraft to remote bases in Texas and other states, returning those ferry pilots to home base following a delivery was a challenge. The lucky ones who could use commercial airline service found that service severely cut back. Many of the DC-3 airliners operating before the war had been commandeered by the military. Ferrying deliveries often were made to out-of-the-way places. For pilots trying to get back to base, taking trains and buses was too time consuming. How could the Ferrying Division best establish a quick turnaround and rapid return of these ferry pilots?

Those commandeered DC-3s, known in the AAF as C-47s, became the backbone of the ferry pilot retrieval service—a military airline set up by ATC. The male ferry pilots nicknamed it SNAFU Airlines—situation normal, all fouled up. It also was known as FUBAR and TARFU.[2] The alphabet soup denoted the colorful, if irreverent, names the men gave to the pilot retrieval service.

Airline pilots—Reservists who flew C-47s/DC-3s before being called to active duty—went to work immediately ferrying the big airplanes and later flew them to pick up groups of ferry pilots. These experienced pilots needed copilots and they got them from the ranks of the ferry pilots who were ready to transition.

The FD/ATC returned planeloads of ferry pilots to base in the shortest time possible while also giving on-the-job transition into bigger, heavier, high-horsepower transports to pilots trying to qualify for pursuit school. Both the male and female ferry pilots were part of this transition plan. They learned to fly the aircraft like the C-47 and the B-25 by serving as copilot for experienced pilots.

"Not many of the male ferry pilots—at least at Long Beach—had ever flown them either," Critchell said. "They began to assign a third copilot, male or female, to build experience. When they sent two copilots along, you prayed you got some time in the right seat. The planes went all over the west, so we were honest-to-goodness copilots. We got the experience of flying a more complex aircraft. We got to fly on all of the ferrying routes, not just the routes we used when we ferried the BTs. This meant we got to fly the routes over mountainous terrain.

"I was very fortunate to be assigned early to these kind of flights because they helped me move up. First, I got checked out in an A-24, and then the A-25 Curtiss SB2-C. Before long, I was being assigned as copilot on a C-47, or a B-25, sometimes even a B-17. These planes were going out to pick up pilots who had delivered aircraft to modification centers around California—Victorville, Daggett, wherever, within 200 to 300 miles of Long Beach. Ground transportation to get pilots home was really sub-standard. They used all of us—the male pilots who had never flown these heavier aircraft and the WASP."

Anna Flynn, 43-6, echoes the use of SNAFU flights for WASP training at Romulus as well, recalling a particular C-47 flight:

"They kept a pretty close watch on flight records at Romulus .... 'Miss Flynn, your file shows you lack four hours of twin-engine copilot time.'

"I told them I'd get around to it. They got around to it for me. 'There's a SNAFU going out this afternoon and you'll be on it as copilot,' I was told." She ended up making the entire flight to Akron and Chicago and back flying as pilot-in-chommand in the left seat, not the right, or copilot's seat.[3]

And, finally, there was Pursuit School.

"General Tunner and the ATC set up Pursuit School so that they could teach several hundred civilian male pilots plus a hundred or so women how to fly pursuits. That's why we were all getting an opportunity to fly heavier aircraft," Iris says. "It was transition through on-the-job training. I think the people of Wilmington were using C-60s, and we had the C-47s and the B-25s at Long Beach. We picked up a nice lot of copiloting experience. In fall of '43, I was put into the sequence, got checked out in the C-47 and was able to go on orders. Then I was assigned pilot-in-command of a C-47 with Dorothy Kocher, class 43-4, as my copilot."

At pursuit school, fledgling pursuit ferry pilots would be trained over the span of roughly four weeks to fly all four of the single-engine pursuits. Ground school training would include the systems and operations of those aircraft. The ferrying bases would send a new class of trainees every couple of weeks.

"Ruth Thompson and I were assigned from Long Beach. We cleared base on December 9 and reported to Palm Springs December 10. When we arrived, there were other WASP there from Dallas, Romulus, and Wilmington.

"We got our lead-in training in the BC-1—which is like an AT-6—by flying off of a sandy dirt strip that was near the big runway. We flew in the backseat with the seat lowered. That way you can't see over the nose. They had us fly the approaches with the flaps up at the approach speeds of a pursuit, so that we began to get the speed relationships. By substituting the BC-1, we got the feel of a pursuit on final approach. That was neat training. That was the dual instruction that made it possible for us to work up to the point that we could fly the pursuits—solo—because all four pursuits we would train on were single-seat aircraft.[4]

"They chose the wide-gear P-47 to be the first pursuit that we flew. It was a tough, remarkable airplane with that big radial engine,[5] so it was very hard to see over or around the nose." (While taxiing the aircraft, the pilots had to make "S" curves, using the rudders to turn the aircraft from one side to the other in order to see where they were going.)

"But we'd been practicing in the BC-1 and we had made the transition. We flew the P-47 through a set of lessons—transition phases of learning the airplane—then we were finished on the P-47 and assigned to the next airplane, which was the P-40."

Seeing over the nose of the P-40 was equally difficult and the landing gear was much narrower. The front wheels were closer together and it was more prone to ground looping.

"The P-39 was a big shift," says Iris. "This was a nose wheel airplane. We had flown no nose wheel airplanes up until then. Everything was tail wheel, so we had that transition to make. Then we moved on to the P-51 and back to a taildragger.

"The P-51, of course, did everything and was a pretty exciting airplane. But it was critical to understand the coolant systems in the engines in these airplanes. The P-47 had the big air-cooled radial engine, but the other three were equipped with inline, liquid cooled engines. If they lost their fluid, you would have to bail out because they'd catch on fire rather quickly.

"We had very good training from the factory representatives—ground school and planned flight training—in all four aircraft."

Two original WAFS, Helen Richards and Dorothy Scott, were sent to Dallas with Nancy Love, January 1, 1943. There, they became good friends, but then Helen was sent to Palm Springs in September with fifteen other WASP to transition into the C-47 and possibly begin pursuit training.[6] When the latter plan was scrapped, Helen was sent to Long Beach. But the two friends were reunited at the first pursuit class scheduled at Palm Springs. They were roommates and planned to go horseback riding the afternoon of December 3. It was not to be.

That day, Dorothy and her instructor were up in a BC-1 trainer. They were on final approach to land. A student pilot flying above and behind Dorothy in a much faster P-39 also was on final approach. Descending, he overtook and came down on top of the slower trainer. The tower, when it realized its mistake, tried to warn them, but it was too late. The airplanes collided mid-air and both crashed. All three pilots died. The blame was laid on the tower, NOT the pilots. Nancy Love had lost the second woman from her original squadron of WAFS.

Devastated at the loss of her friend, Richards threw herself into learning to fly pursuit. On December 24, she checked out in the P-51 Mustang, earning the coveted "3p" classification. In 1944 Richards, stationed in Long Beach, would log 110 hours and 35 minutes ferrying P-51s cross-country, the most hours she flew in any one airplane during the war.

Dallas pilot Cappy Vail, 43-2, a member of the pursuit class that graduated January 10, recalls, "I had a wonderful time. I was anxious to fly a P-40, because Art [her fiancé] was flying those overseas.[7]

"Nobody had ever told me about the torque on that engine. You have to practically stand on one rudder because the plane wants to reverse itself and go the opposite way. So you have to keep it straight by using the rudder. That was a surprise. It took a lot more strength to fly the P-40 than anything I had flown before. But I loved it because Art did."

In December, before the first group of women graduated from pursuit school—in fact before they had flown the P-51, the last of the four pursuits in which they would check out—several were taken to Long Beach to take part in a special delivery. Long Beach needed to move a bunch of P-40s because the airplane parking area between the runways was overloaded.

Dallas WASP Betty Whitlow has vivid memories of that delivery. It was a beautiful experience until she touched down on the runway at the Oakland Naval Air Station. "I was cleared by the tower and made a beautiful landing. A split second later, a little plane is crossing the runway, right in my path.

"P-40s are taildraggers, so I cannot see anything. I'm going 100 to 120 miles an hour. And there are these two fellows in this little plane. I got so close I could see their expressions of shock.

"I put on the brakes. Nothing else I could do. At that same instant, the tower comes on the radio and says, 'P-40 on the runway ...' Well, I put them on a little too hard, just hard enough that the plane sort of eased up and then nosed over. It bent the propeller!

"This plane was supposed to go out to the Pacific that night. I was furious! I sat there fuming. I turned off switches, and I still was fuming. The tower had cleared me to land, and I made a beautiful landing. I don't know why he was there, but the damage was done. I contacted my base. They never blamed me for anything.

"It was my first ferry trip in a pursuit! That's what infuriated me so. And I was landing at a Naval Air Station!"[8]

Four more of Nancy Love's "Originals" roomed together in Palm Springs, arriving December 10, 1943. Nancy Batson and Gertrude Meserve shared one room in the barracks—two iron cots and a partition for a wall—and Helen McGilvery and Barbara Donahue, now the WASP squadron commander in Romulus, bunked right next door.

Immediately after graduation in Palm Springs and delivering a high-priority P-51 to Newark, Batson, McGilvery, and Meserve joined the rotation of Wilmington-based pilots up to the Republic factory in Farmingdale, New York, where the P-47s were built. And as more and more pursuit-trained pilots came back from pursuit school, squadron leader Betty Gillies added them to the list. As her contingent of pursuit ferry pilots grew, she could send eight at a time for two weeks of TDY and still have pilots in Wilmington to ferry other aircraft as needed.

And one more Original WAFS, Sis Bernheim, would graduate from pursuit school in July. She, too, took her place ferrying mostly P-47s from the Republic factory on Long Island to Newark or to and from the modification center in Evansville.[9]

"The training at pursuit school was outstanding," Iris says. "After graduation, we all had immediate orders, to go to Long Beach January 10—the women and the men. They had so many new P-51s there, all of us were sent to make deliveries. For those of us stationed at Long Beach, that was the beginning of a constant flow, moving P-51s to Newark."

Graduating in the first two pursuit classes were Betty Jane Bachman, Lewise Coleman, Iris Cummings, Norma Jane Emerson, Carole Fillmore, Ruth Thompson, Betty Whitlow, and Cappy Vail from 43-2; Sidney Miller from 43-1; and WAFS Nancy Batson, Helen McGilvery, Gertrude Meserve, and Helen Richards.

More WASP arrived at Palm Springs for pursuit training January through March 1944. Graduates from the Guinea Pig Class, 43-1, were Mary Lou Colbert, Byrd Granger, Marjorie Ketcham, and Jane Straughan. From 43-2: Virginia Alleman, Ruth Dailey, Kay Gott, Ross Kary, Jary Johnson, Ruth Grimm Trees, and Mary Trotman. And from the original WAFS, Barbara Donahue and Florene Miller.[10]

Seven pursuit classes graduated in Palm Springs—the last one on April 1, 1944. The school was then moved to Brownsville, Texas, and that is where all subsequent WASP pursuit transitions took place.

Jean Landis, 43-4, from Long Beach was one of the ninety-six WASP ferry pilots to graduate from the pursuit school in Brownsville.[11]

"For several months, my trips out of Long Beach were delivering BTs and an occasional AT-6," Jean recalls. "Good weather, good planes, good friends, good food, no sweat. Then in June of '44, I was sent to Brownsville to train in fighters.

"Living conditions were okay. Our instructors were great and the flying was fantastic. We spent a lot of time flying from the backseat of an AT-6 as they felt this position best paralleled the way it will feel when you got into a fighter, especially the P-51. We also spent a lot of time sitting in the various fighters studying the instrument panel and going through various emergency procedures.

"We took a blindfold test in each aircraft. Sitting in the cockpit blindfolded, your instructor, who stood on the wing, called out 'point to your air speed indicator'—'reach for your left wing gas tank control'—'reach for your flaps.' Finally, after many hours of this preliminary training, we were considered ready to go.

"We were scared to death—the guys too, not just us gals. One of the guys started to sing 'When the roll is called up yonder I'll be there.' We all chimed in. Slips bearing the numbers P-39, P-40, P-47, and 51, were put in a hat and we each drew the pursuit we were to fly for the very first time.

"I drew the P-51, the love of my life. How lucky can you get!

"In the cockpit, waiting to take off, I applied and held the brakes, revved up RPMs, watching the oil pressure and oil temperature gauges as I did. Then, I got the word 'ready for takeoff.'

"Full throttle, off with the brakes. I never felt such pressure against my back. The power and the noise were unbelievable. Thrilling! Exciting! I was off. I upped the flaps, upped the gear, in seconds I was at 1000 feet, then 2000 feet, where I was told to practice simulated landings. That's where you get into a nice glide path, let down the gear, lower the flaps, reduce your air speed to about a hundred and ten—like you're coming in over the fence, almost ready to touch down. Then give it the gas, up the gear and the flaps, get proper altitude, and do it all over again, and again, and again.

"All was well until the red warning light came on indicating my landing gear was not down and locked. Immediately, I went through the emergency procedures. I climbed then dived down, pulled back on the stick to pull quickly out of the dive in hopes of locking the gear.

"Nothing worked. So I called in that I had a problem.

"'Come on in, let us see,' they said.

"I was so timid the first time, I didn't go low enough. They couldn't see if my gear was down or not. So I dove on the tower a second time. This time, I came in so low I could almost see the whites of their eyes. 'Yes,' they said, 'the gear is down, but we can't tell if it's locked.'

"I had to go back up and fly around while they cleared the field for me to make an emergency landing. My instructions were, 'Come in softly,'—whatever that means—'gently touch the wheels to the runway. If they hold, taxi on in. If they don't, give her the gun, climb, get to altitude and circle until you get further instructions.'

"As I headed for the runway, here came the hash wagon and the fire trucks. Remember, this is my first landing in the P-51. I came in as gently and softly as I could. Touched the wheels to the runway. They held. I taxied in.

"My instructor started in on me. His first words were, 'Well, what did you do wrong?'

"'I think I did everything properly, Sir!'

"I got out and he got in and took off. The last thing I saw, he was all over the sky, diving and sharply pulling up, trying to get the wheels down and locked and the light to go off. I wanted to shout, 'I told you so.'"

Chapter 21

Militarization

Congressman John Costello of California had introduced a bill in Congress in fall 1943 calling for the militarization of the WASP. The bill went to the House Committee on Military Affairs for study. Subsequently it was amended to include the appointment of female trainees (the women at Avenger Field) as aviation cadets. The amended bill to militarize the WASP was submitted to Congress in February 1944.

Though directly impacted by the outcome of the bill, the Ferrying Division was never consulted. The Ferrying Division favored militarization of the women pilots, but not under Jackie Cochran.[1]

Why militarize the women pilots?

Their lack of military status was a form of discrimination against them. Military status would give them military insurance, death benefits, hospitalization, and pensions. And continuity of their service would be ensured.[2]

The possibility of taking the women pilots into the WAAC and later the WAC was brought up four times—all to no avail. The most serious effort occurred late in 1942, just before Class 43-2 entered training and the WAFS were broken up into four squadrons.

In early December, Arnold called Cochran back to Washington from Fort Worth to meet with Col. Oveta Culp Hobby, Director of the Women's Army Auxiliary Corps (WAAC),[3] about the possibility of putting the women pilots under the WAAC. Cochran related the conversation in an interview years later:[4]

"I got a chip on my shoulder the minute I walked in. ... Her office was more important than Arnold's, larger. ... I'm not sure it wasn't larger than General Marshall's."

Hobby kept Cochran waiting. Once they were seated face-to-face, Hobby tried to engage Cochran in ice-breaking chitchat. Receiving no response, she went on with the business at hand. "Now when we take over your outfit, naturally I expect you to come in and be their immediate superior."

Cochran informed Hobby that she, herself, had not been consulted, nor had "the girls," nor, to her knowledge, had General Arnold.

Hobby, according to Cochran, replied, "Well, that's unimportant. I think it will be a great asset to them and to the WAFS."

"You've bitched up your own outfit. You are not going to bitch up mine," was Cochran's reply. And that was the end of what turned out to be the most serious attempt concerning the potential avenue for militarizing the women pilots.

Then on June 25, 1943, with Cochran on the verge of being named Director of Women Pilots, she and Colonel Hobby had a second conversation. Col. R. H. Carmichael witnessed the conversation: [5]

Cochran: "Women pilots are very temperamental. [Their] administration should be handled by people who do understand them—not by people who have no appreciation of the type work they are doing."

Hobby: "I don't know what you mean by administration."

Cochran: "I mean people over them who understand their problems."

Hobby: "I don't think it is very likely that you would have anybody over them who didn't."

Cochran: "Yes, I think so."

General Arnold put the matter on hold for sixty to ninety days.

The WAAC became the militarized Women's Army Corps (WAC) on July 1, 1943. On July 19, 1943, ATC's Gen. C.R. Smith asked Arnold, again, to consider a merger of the WAFS with the WAC. Action was needed and many AAF officials saw the advantage of militarizing the women pilots, but Cochran and Arnold opposed this, preferring to (if possible) commission the women pilots directly into the Army Air Forces.[6]

General Tunner tried one last time in November 1943. He personally urged Gen. H. A. Craig to put the women ferry pilots into the WAC with commissions. Following his conversation with General Craig, Tunner had Nancy Love draft a letter for his signature, to be sent to Colonel Hobby. In the letter, Mrs. Love laid out the various adjustments that would be necessary for the women ferry pilots to become part of the WAC, and she requested that Hobby give her an opinion of each of the proposed adjustments. WASP could be as young as eighteen-and-a-half but WACs had to be twenty-one; a WASP could have dependent children under fourteen, a WAC could not; and a WASP who failed to qualify as a pilot could be returned to civilian life, whereas a WAC could not.[7]

The letter never reached Colonel Hobby. Captain Marx writes that it was stopped and held by A-1 of Air Staff and pigeonholed.[8] (A-1 is the staff officer concerned with policies and plans relating to personnel.[9]) Cochran, in her early 1945 rebuttal to Captain Marx's report on the "Women Pilots in the Ferrying Division," wrote, "General Tunner again tried to press the matter and Mrs. Love wrote a letter to Mrs. Hobby *which had to be stopped* and pigeonholed by A-1 of the Air Staff."[10]

This appears to indicate that Cochran was made aware of Nancy Love's letter by the A-1. Cochran was stationed right there in the Pentagon and, as part of the Office of the Assistant Chief of Air Staff, was privy to what went on. It may have been Cochran who stopped the letter from reaching Colonel Hobby.

Cochran goes on to state that her reason for not wanting the WASP in the WAC was that "placing a flying organization that worked as an integral part of the Air Forces under a ground organization headed up outside the Air Forces would have added complications rather than otherwise."[11]

By then, the Arnold/Cochran plan for militarization already had moved too far forward to go back.

By the beginning of 1944, the entire tenor of the war had changed. The Allies had turned that all-important corner and were now on a path that would lead to victory. Roadblocks to WASP militarization began to crop up on several fronts.

Arnold had overestimated the number of pilots he was going to need to win the war. On January 15, 1944, he terminated the WTS program for training flying personnel and the AAF began to cut back on its own pilot training program. Men who couldn't qualify for pilot slots would be given shots at navigator, bombardier, technician, or other training. Failure there would land them in the Ground or Services Forces.[12]

A year earlier, when Nancy Love's WAFS had begun to stretch their wings and ferry airplanes, there was a demand for more pilots. Now the availability of pilots had done a 180. Rather than a dearth of male pilots, there was a surplus. As a result, the WASP became increasingly aware that their presence was no longer looked on as the saving grace it had been.

Nancy Love held periodic conferences with her squadron leaders and their assistants. To plan for the New Year, she called the group together in Cincinnati on January 5, 1944. Present were: Barbara Erickson and Bea Medes from Long Beach; Delphine Bohn and Avanell Pinckley,

Dallas; Lenore McElroy and Margaret Ann Hamilton, Romulus; and Betty Gillies and Esther Rathsfelder, Wilmington. Romulus squadron leader Barbara Donahue was at Pursuit School, so third-in-command Hamilton accompanied second-in-command McElroy.

Majors Gordon Rust, Fry, and Smith from the Ferrying Division met with them on January 6 to talk about upcoming training. Betty Gillies noted, "It appears that the ships we are most needed to fly are P-47s, 51s, 39s, 40s, P-38s, A-20s and A-30s[13]. So we are setting our training program with that in mind."

This information, of course, dovetailed with the news coming out of Palm Springs. As the WASP ferry pilot leaders were meeting, some of their women were becoming the first women to graduate from Pursuit School.

Betty, who recorded every day's happenings in her diary, wrote:

> The subject of militarization has been painfully avoided! Off the record, B.J., Delphine, Esther and I are going to go to Washington and see what's what! We got orders for B.J. and Delphine to proceed to NCAAB "to coordinate WASP affairs." We are leaving here tomorrow a.m. planning to delay in Washington en-route.[14]

On January 7, 1944, they saw Jackie Cochran, "to feel her out," Betty writes. "Spent about two hours in her office and got nowhere. But it was an interesting experiment!"

On January 8, they visited the Navy. The Navy was interested in using women ferry pilots. Then, back at the Pentagon, Betty met with Brig. Gen. William E. Hall, Deputy Chief of Air Staff.

> I asked him if it were possible for us to ask for and receive a commission in the AUS [Army of the United States] and service pilot rating.

"By golly," he said, "if you can, that would solve all our problems."

In October, General Hall and Air Staff had sought an amendment from AAF General Staff that would make it possible to guarantee women aviation cadets the same privileges and benefits as the men.[15] The General Staff turned down the proposal. However, General Arnold opposed the General Staff's decision and on January 5, 1944, said so publicly and recommended amendments that would make possible the women pilots' direct commission in the AAF.[16]

The four WASP spent January 9 in Wilmington:

> working up a proposed regulation permitting us to apply for a commission in the Army as service pilots. We checked all the Army regulations and AAF regulations and can find no snags. – Bud came by that night, took us all out to dinner at the DuPont Hotel. [Bud was Betty's husband.]

On January 10, three of them—Betty, Delphine, and B.J.—went back to Washington to see General Hall, who "seems to approve of our ideas" but needed more time to work on the proposal. Delphine and B.J. called their base commanders to get permission to remain a couple of days longer and Betty returned to her duties at Wilmington that evening.[17]

In a memo dated January 11, 1944, General Hall asked Air Staff Personnel to:

> look into the legality of commissioning women pilots directly into the Army on the basis of their qualification as service pilots, stating that if such legal basis could be found, General Arnold would be highly pleased because it would be the answer to militarizing the WASP and at the same time keeping them out of the WAC.

Unfortunately, the reply two days later stated:

> the authority required extended only to men and could "not be regarded as authority for commissioning women as officers in the Army of the United States."[18]

***

Esther Manning (Rathsfelder) also kept a diary and recorded her thoughts and revelatory personal impressions of those days in Cincinnati and Washington D.C. She and Betty Gillies took the overnight train to Cincinnati, arriving at 9:30 the morning of January 4.

> January 4, Tuesday.
>
> Spent the afternoon with Nancy Love at Headquarters. Learned I have to have solo time on Class III transition. How the base will do it with one beaten up airplane I don't know. We should stop the girls who don't want pursuits at Class II transition. That was Betty's idea to begin with. So far nothing new is startling her.
>
> January 5, Wednesday.
>
> Long, long day of meeting and problem discussing, but it's swell to be seeing Delphine and B.J. again—as natural as if we'd parted yesterday. If only Donnie could have been here, but she's at Palm Springs at pursuit school.
>
> This morning was mostly a rehash of things we've been discussing at Wilmington for weeks and gripes from the other bases, which made us feel disgustingly smug. However, this afternoon we acquired [information] on allocations of airplanes, which helps clarify the picture of what each base had to offer. Also informed we are entitled to sufficient qualified office help. Also can have four responsible people on each base and only one has to be there at a time. [This was the result of AAF Reg. 40-8, issued December 21, 1943.]
>
> January 6, Thursday.
>
> Today ended the conference. Not particularly eventful. We may get a chance at the weather course at St. Joe. And the futures of the gals now at Pursuit School seem to be the pursuits, A-20s, A-30s, and P-38s—a good deal.
>
> B.J. and Delphine got new orders so they can proceed to Washington with Betty and me tomorrow and we will see what trouble

we can stir up for J. Cochran there. Going to try the angle of having us commissioned as service pilots through our own bases as the men civilian pilots are. I should think the war department and a couple of key congressmen might be the spots, but we shall see what the consensus of opinion will be.

Nancy is certainly a beautiful woman—lovely to look at, in all truth. She is truly on the spot in this. As she said tonight, she can't quit even if militarized under J.C. and still make a decent return for all the good things ATC has done for her. And, of course, being a lady she cannot meet J.C. on her own ground.

January 7, Friday.

Today I met the great J. Cochran. She is prettier and much less hard looking than I thought she would be, and has undoubtedly much improved in manner since her hairdressing days. She has acquired a certain charm but her terrific drive doesn't give it much chance to operate. She unquestionably has a genius for getting what she wants.

I can only say that she is terrifying—the kind of a person with whom you could never hope to do business. She will have her own way—no one's suggestions will be heeded. She must do anything and everything herself and get full credit for it. The only answer to her must be to defeat her completely. Tonight, over dinner, it appeared as if we might have the answer.

Why can't we make our application for a Service Pilots commission through Air Corps channels, as the men Civilian Pilots do? B.J. is going to try it and we will see what happens.

January 8, Saturday.

Back to the base tonight after a day mostly sitting and waiting to see Mrs. Roosevelt—which we didn't. However did see the Navy ... and they would be glad to have us if we can't get what we want in the Army. Also, Betty saw General Hall, way up on General Arnold's staff, and presented the service pilot idea to him.

> Apparently they had never thought of it, because he was most enthusiastic. Betty is to go back to see him Monday.
>
> January 10, Monday.
>
> Betty ran down to Washington today, and it looks very much as if our plan would be accepted. Nancy, however, got wind of our being in Washington and is very aroused. Thinks we'll spoil everything, which I certainly doubt we could, considering the sickly measures she has taken. However, she is naturally timid—too much so to do battle with a woman of Cochran's type.
>
> So, soon I shall be faced with the terrible decision about militarization. I'm sure I can't accept it, but it will sure hurt not to.[19]

***

In her memoir, Delphine Bohn shares her recollections of contacts she made during that stay in Washington. Most impressive was a visit to "Mr. Sam," the venerable Speaker of the House Sam Rayburn (a fellow Texan). She also went to see her Congressman, Rep. Eugene Worley, and a neighbor from Amarillo, Under Secretary of Agriculture Grover Hill. Finally, she called on Gen. Brehon B. Somervell, Commanding General, Army Services Forces. She notes that Somervell "had under his command military personnel of the Women's Air Corps category."

But by then, she and B.J. had established very high profiles in Washington. When Nancy Love found out what her trusted lieutenants were up to, she knew she had to put a stop to it. The Arnold-Cochran plan to militarize the WASP was nearly ready. Too much was at stake.

Delphine writes:

> Someone, helpfully, notified Cincinnati and Nancy Love. Since we were registered at the Hay-Adams Hotel and the whole of the military community had become familiarized with our many visits about town, she had no difficulty locating us.

Nancy telephoned them in Washington and, according to Delphine, "very gently, but very firmly, ordered us back to our bases."[20]

The trouble really began when B.J. and Delphine got back to their respective bases. "Sadly," Delphine writes, "in our complete lack of thoughtfulness, we had caused our admirable base commanders trouble."

The culprits were reprimanded for their "illegal use" of Transportation Requests. They had to repay the cost of their unauthorized fares, which, says Delphine, they intended to do anyway. "But these reputable officers had to accept from headquarters a share in our reprimands. For a week or so following our return, our commanders could not decide whether to smile, laugh or frown heavily whenever they ran into us on base. We attempted to avoid them."[21]

"Most of the women ferry pilots didn't want to be militarized under Jackie Cochran," Barbara London said, in a 2004 interview with the author, which follows here:[22]

> "We were interested in militarization as individuals. Militarization would provide the benefits we knew we needed. But we didn't fit anywhere as a group. That had already been tried when they wanted to put us in the WACs. So I applied for a commission to become a Service Pilot. I had the qualifications. Using my initials, B.J. Erickson, they wouldn't know if I was male or female.
>
> "It was a good idea, it was logical. The women ferry pilots were for it. But it fouled up Jackie Cochran's idea to make us a separate group under her command. She wanted equal status with Oveta Hobby and the other women commanders of military units —the SPARs, the WAVES, the Women Marines. She needed the thousand-member WASP for that. Jackie's influence was unbeatable—her wealthy husband, both Roosevelts, Arnold. And Nancy thought she could bring it off."

So ended the only grassroots attempt by the women ferry pilots to militarize.

***

The news that followed swiftly on the heels of these events signaled a positive shift in the progress of the war. The week ending February 26, 1944, long-range P-51 Mustangs escorted U.S. heavy bombers on 3,800 sorties over Germany shooting down 517 German fighters, killing 225 air crew members and wounding 141—almost 10 percent of the skilled airmen in Germany.[23] On March 4, P-38 Lightnings escorted B-17s over Berlin in the first U.S. daylight-bombing raid since the Battle of Berlin began November 18, 1943.[24]

The revised bill including WASP militarization was introduced February 17, 1944, as H.R. 4219. The Secretary of War recommended its passage. An identical bill, SB-1810, was presented in the Senate on March 24, 1944.[25]

What the Costello bill said (as paraphrased by the author) was:

For the duration of the war, that women be commissioned as flight officers or aviation students in accordance with existing regulations. No woman would be appointed to a grade above Colonel and there would be not more than one officer of that grade. Female flight cadets, upon successful completion of the prescribed course of training, would be commissioned as second lieutenants or appointed as flight officers. All commissioned women would receive the same pay and allowances as male members of the Army of the United States and that they would be entitled to the same rights, privileges and benefits according to their rank, grade and length of service.

Introduction of the WASP bill followed close on the heels of the closing of the AAF's pilot training program. Congressmen began to receive letters protesting the militarization and instruction of women pilots. Because of the women pilots, trained male pilots claimed they were being put in the "walking Army." Letter writers were members of the American Legion and other veterans' organizations as well as mothers of boys transferred from aviation cadet training to the infantry. Their tone was angry and full of resentment.

Congress changed the complexion of the hearings when it urged the House Civil Service Committee to study of the question of WASP training and utilization. Georgia Congressman Robert J. Ramspeck—a former deputy U.S. Marshal—was the chair and the committee forever after bore his name. The WASP were under Civil Service, therefore public funds were being expended on their program—a program that Congress knew next to nothing about.

Jackie Cochran, Ethel Sheehy, and Nancy Love, all dressed in their new Santiago blue WASP uniforms, sat in attendance as General Arnold testified about the WASP before a friendly House Armed Services Committee on March 22, 1944. Fearing that the entire women's program was in jeopardy, given the uproar and the bad press, Nancy buried any feelings about Cochran. They presented a united front.[26]

General Arnold explained that the Army was short more than 200,000 men and the Air Forces needed to return 36,000 men—scheduled for aircrew training—to the Ground and Service Forces. He urged the passage of the WASP bill, citing the shortage of manpower in the Army and in key industries. He wanted the authority to use the WASP as full military personnel for non-combat service[27]—to replace men with women wherever possible.

Following the open session, the committee held a closed session in which General Arnold was the only witness. At no point was the Air Transport Command asked for its input on the women ferry pilots who flew for the ATC's Ferrying Division.[28]

When the meeting concluded, the Military Affairs Committee reported the WASP bill favorably out of committee, recommending passage by Congress and it was sent to the House Rules Committee awaiting a place on the House calendar. Up until then, Congress had given Hap Arnold anything he asked for, but now things changed. Instead of quickly passing the bill that General Arnold personally asked for and fully supported, the politicians in Congress listened, instead, to increasingly loud and frequent outcries in the press and from pressure groups. The press had a field day at the expense of the women.

While some of the mainline newspapers—the *New York Times* and *Herald Tribune* and *Boston Globe* leading the way—were supportive of the WASP, others took the side of the civilian men pilots, writes author Sally VanWagenen Keil, *Time* magazine among them.

"The WASP program was called everything from a 'blunder,' to a 'fast-play' and a 'racket.'" One article was headlined "Wanted Female-Impersonators," while another referred to "Jackie Cochran's glamour girls." "How about some of these 35-hour female wonders swapping their flying togs for nurses' uniforms?" said another. That spring of 1944, WASP ferry pilots out delivering aircraft around the country began to notice the change of attitude toward them, and it often was manifested by abusive language and incidents. A backlash of public opinion against women had begun in wartime America.[29]

Scholarship that has come out of the Women's Movement suggests that the rationale behind this is more complicated that it appears. Male pilots complained to Congress that the WASP ferry pilots received preferential treatment, were given more opportunity to fly, that they were flying more frequently than the men and had better assignments. They claimed men sat on the ground while the WASP were kept busy.[30]

AAF Headquarters asked General George for comments. George, reading the tenor of the times and acutely aware of the gravity of the situation, asked General Tunner to compile a complete report concerning the WASP performance on the job in the Ferrying Division. "Prepare for me immediately," he said. His final words carried an urgency. *"Make it thorough and make it good."*[31]

General Tunner sought input from the commanders of his Ferrying Groups and from the commander of Pursuit School newly moved to Brownsville, Texas. All had daily dealings with the WASP ferry pilots. He knew he would get an accurate report from them.[32]

Resentment did exist among a minority of the male pilots, Tunner's commanders told him, and that minority was quite outspoken. But the supervisory and operational personnel, whose sole interest was the efficient operation of the ferrying mission, did not share this resentment. "Those personnel are familiar with the pilot shortage in the Ferrying Division and appreciate the WASPs' contribution."[33]

General Tunner addressed the charges of favoritism. "The WASP flying time has consistently averaged slightly less than the flying time of male pilots," he wrote. WASP pilots took their turn. Normally, any pilot in the Ferrying Division, including the WASP, who had returned to his or her base was placed at the bottom of a roster consistent with the type of aircraft he/she was qualified to fly, and ferrying missions were assigned to the individual who had been on base the longest. In the spring of 1944, the Ferrying Division had been extremely busy with domestic operations and all pilots, regardless of sex, had been utilized without anyone waiting an undue length of time for assignment.

Tunner's conclusion: "WASP have assisted in the performance of the Ferrying Division's mission at a time when assistance was not available from any other source. In this respect, it must be noted that during the past three-month period all leaves to Ferrying Division pilots were cancelled due to a pilot shortage."[34]

An examination of the report of the Ramspeck Committee and of the debate on the bill in Congress gives no indication that this material furnished to General George was utilized by the AAF or ever presented to the members of Congress.[35]

George and Tunner were confident of their own people, did their part, and proved the naysayers wrong. Their effort came to naught.

In the meantime, the Ramspeck Committee made it clear that its greatest concern was the contemplated recruiting of inexperienced personnel, not the use of women as pilots.[36]

The bill did not go to the floor of the House until June.

Chapter 22

Officer Training School

The AAF considered WASP militarization a shoo-in once the Costello Bill went to Congress in February 1944. In March, General George asked the Ferrying Division bases to introduce close-order drill three hours a week for all WASP, "with particular emphasis on snappiness and advanced maneuvers."[1]

Getting the WASP prepared for commissions in the AAF was uppermost in Cochran's plans. All commands using WASP received a letter on March 27, 1944, stating: "As Women Airforce Service Pilots are becoming an integral part of the Army Air Forces it is desired that all WASP personnel be given training similar to that for Army Air Forces officers." General Barney W. Giles signed the letter for General Arnold.[2]

The WASP were going back to school.

The School of Applied Tactics (better known as Officer Training School or OTS) was located in Orlando, Florida. The first class reported on April 19, 1944. Class duration was to be four weeks. To be eligible, a WASP had to have no fewer than ninety days' operational duty and, when possible, they were to be selected in order of length of service.

Twenty-four women were assigned to the first class, including Nancy Love and the four squadron commanders from the ferrying bases—original WAFS Barbara (B.J.) Erickson, Delphine Bohn, Betty Gillies, and Barbara Donahue.

Giving pursuit ferry pilots TDY (temporary duty) to attend OTS put a strain on the Ferrying Division. As General Tunner pointed out in his letter to General George, the Ferry Command was so swamped with deliveries during the spring of 1944 that no pilot leaves were authorized.[3] The exception was the WASP sent to OTS. On top of that, the Congressional controversy was in full swing, and the women's sojourn in Florida began to provide more negative fodder for the press and other detractors. With the Ramspeck Congressional Committee examining the program in minute detail, removing several key women from their on-duty jobs and sending them for training unrelated to ferrying seemed ill advised. And another twenty-five were scheduled to report the first Wednesday in May and every first and third Wednesday of the month thereafter.

The curriculum consisted of military discipline, customs and courtesies, chemical warfare, and organization of the army and staff procedures. For some it was a much needed respite—a vacation away from the stresses of ferrying, something they hadn't had in a long time. Others considered it a boondoggle—a waste of time when they were needed elsewhere.

Byrd Granger (43-1) took a thoughtful tack and writes, "Those who might have gone through the special WASP course, but willfully and successfully evaded it, missed an experience which places its graduates in a unique position ... being among the few American women who attended the Army Air Forces School of Applied Tactics, a rare privilege."[4]

For class, the women wore their dress uniforms, but with collars open, no ties—a concession to the Florida heat. They marched by twos to class, spent evenings studying in the library, and had Sundays free.

Whatever the individual and collective benefits of OTS, the timing turned out to be disastrous. While the first class was in training in

Orlando, and as the second was reporting, the May 8 issue of the weekly national news magazine *Time* carried one of numerous articles about militarization that appeared in the country's magazines and newspapers that month. The article was headlined "Battle of the Sexes."

Already the Army Air Forces was giving its women pilots officer training before the bill had even reached the House, the article charged. "How they jumped the legislative gun was an official mystery. But the news sparked new outbursts in the stalemated battle of the sexes over who is to ferry the Army's transports." Then the article alluded to the "flood of indignant letters to Congressmen." And—as if to add insult to injury—it also stated that women pilots were grounded several days out of the month, a statement that was completely false as per evidence obtained by the Surgeon General of the Ferrying Division.[5] Capt. Nels O. Monserud, Air Surgeon's Office on WASP Personnel at Avenger Field, concluded, "Menstruation is not a handicap in selected candidates to prevent a woman from fulfilling her job as a pilot of military aircraft."[6]

Ruth Roberts (43-6), stationed at Romulus and who was attending OTS when the article came out, felt the article was so blatantly unfair that she wrote a letter addressed to the editors of *Time* in which she staunchly refuted the article and its claims.[7]

"Dear Sirs," she wrote, "*Time* states, 'The Army Air Forces has been surreptitiously giving officer training to women civilians.' ... We have been ordered to attend Officer Training at Orlando," she went on, "though [none] has ever asked to be commissioned."

Contrary to claims of deficiency in hours, many of the women pilots, she pointed out, held "C.A.A. commercial, instructors, and instrument ratings, and A&E licenses." And, "the original WAFS were required to have 500 hours, and entered the service on the same status as the men without additional training."

From personal experience, she informed the magazine and its readers that in her time at Romulus, "very few women have ever found it necessary to be grounded because of any affliction peculiar to their sex."

She concluded with, "The women flyers of Britain and Russia, who are doing ferrying and transport work, proved sex was no barrier to flying ability long before the women in the United States were accepted by the Air Transport Command. If only 500 women have proved they could relieve 500 or more men for more action flying duties, surely the WASP deserves to be commended—not condemned."

The letter was signed, "Ruth Roberts."

Beginning in April 1944, pursuit flying became *the* major activity for the WASP of the Ferry Command. The pursuits were swift, high-performance aircraft; therefore, flying them carried an extra element of danger. The first WASP ferry pilot to die in a pursuit aircraft happened to be one of the most experienced—original WAFS Evelyn Sharp. Before the program ended, three more women pursuit pilots would die—none of them with Evelyn's experience level. Nebraska educator, pilot, and Ninety-Nines member Diane Bartels has written Evelyn's biography, *Sharpie.*[8]

When twenty-three-year-old Evelyn joined the WAFS in Wilmington in October 1942, she had nearly 3,000 hours. On June 13, 1943, Evelyn, nicknamed "Sharpie," was the second woman to fly a P-51—Nancy Love being the first.

Three Long Beach WAFS—Barbara Erickson, Barbara Towne, and Evelyn—went to Palm Springs to begin transition on the P-51A (A-36) model in June 1943. Evelyn soloed first, but the airplane was needed elsewhere so Erickson and Towne didn't get their first flights until July, after which Towne—four months pregnant—went home. Erickson made her first P-51 delivery on July 24, 1943.

In early August, Evelyn had the opportunity to practice takeoffs and landings in a newer model P-51B. On August 14, she took her first P-51 to Newark.[9] She continued to transition into, and to ferry, all the airplanes available to the Long Beach pilots—both single and twin-engine aircraft of ever-increasing horsepower and complexity. She earned her instrument rating in January 1944. On March 26, she made her first flight in the twin-engine P-38 Lightning. She completed transition and was signed off March 29.

Her orders were to deliver a P-38J to Newark. She left Long Beach for Palm Springs on March 30 and departed Palm Springs March 31. She RONed in Amarillo. On April 1 she headed for Cincinnati Lunken Airport for a meeting with Nancy Love. They would discuss Evelyn's new assignment as the commander of the WASP squadron at the soon-to-open 21st Ferrying Group base in Palm Springs.

On April 2, she left Cincinnati intent on delivering to Newark that afternoon, but weather over the Allegheny Mountains was marginal. Just past Harrisburg, Pennsylvania, she ran into "snow, rain and sleet." She turned around, deciding to RON at Harrisburg's New Cumberland Airport.

The following morning, April 3, she was wheels up at 10:29. Her left engine, belching black smoke, quit. Witnesses said she did not turn toward the populated area to the right. Beacon Hill and radio towers lay dead ahead. Going against the rule, never turn into a dead engine, she chose the path that took her away from the town and away from the hilltop. She began a shallow, climbing bank to the left—uphill, where there were but a few farmhouses. The Lightning yawed and skidded to the left. She tried to put it down in a clear spot near a wooded ravine, but when she pulled back on the control wheel, the airplane stalled and hit the ground in a flat position. The time on the clock inside the cockpit stopped at 10:30.[10]

The WASP known as Sharpie had put many a 1930s wood and fabric aircraft down in an open field during her flying life in Nebraska's farm country, but the P-38 was 17,500 pounds of flying steel and aluminum.[11] Evelyn, number seventeen to be accepted into Nancy Love's original

WAFS squadron, was the third—and final—of Nancy's Originals to die during the war.

Before the WASP were through flying, there would be more deaths among Nancy's ferry pilots—these from the ranks of pilots who came out of Sweetwater and Houston.

As of April, there were 8,394 male ferry pilots in the Ferrying Division.[12] Of that number, 1,767 flew pursuits and the ATC was moving as quickly as possible to get 1,770 more civilian pilots checked out and commissioned.[13] In April, the number of WASP in the Ferry Command reached its high-water mark: 303. Thirty-five of them were pursuit qualified. Already, the push was to advance more women to pursuit. The invasion of Europe was imminent. Though it was a secret, rumors were rampant and a certain tension hung in the air. Hap Arnold needed as many men ready to fly as it was humanly possible to train, and they were needed to fly twin- and four-engine bombers and transport overseas. Their transition was critical to the war effort. Consequently, the number of female ferry pilots who could be transitioned for pursuit freed that many male pilots to be assigned where they were most needed.

This was General Arnold's whole rationale and he tried, valiantly, to sell the WASP to Congress and the American public.

By April 1944, the P-51 Mustang was about to change the war in Europe. The P-47s and P-38s being used to defend B-17s and B-24s on their bombing runs to destroy Germany's war industry did not have the range to go the distance to the targets deep in Germany and return with the bombers—leaving them prey to German fighters. In the early models of the P-51, its impressive speed "fell off" above fifteen thousand feet. "The problem was solved by installing the British Rolls-Royce Merlin engine, which thrived at high altitudes," writes Barrett Tillman in his 2014 release *Forgotten Fifteenth*. "As a weapon, the Mustang was tops. ... The P-51 was one of a handful of game-changers among military aircraft. It probably saved the daylight bombing offensive when relentless attrition threatened to force a change in AAF policy."[14]

The newly minted WASP pursuit pilots were destined to ferry P-51Ds to the docks at Newark, particularly the graduates from Long Beach and Dallas, where Mustangs were built.

On April 1, the seventh pursuit class was the last class to graduate from Palm Springs. Pictured in the very center of twenty-three male graduates is one woman, petite Mary Lou Colbert—destined not only to fly fighters, but for the command job Evelyn Sharp was to have had in Palm Springs as it transitioned into a ferrying base.[15]

Pursuit School was moved to Brownsville, Texas, and fifteen women graduated in the first class there on May 1, 1944: Claire Callaghan, Cornelia Colby, Mary Darling, Ruth Franckling, Ellen Gery, Gretchen Gorman, Margaret Ann Hamilton, Vega Johnson, Margaret Kerr, Florence Lawler, Nadine Ramsey, Barbara Russell, Betty Tackaberry, Martha Wagenseil, and Barbara Willis.

The May 15 class produced thirteen women graduates: Betty Archibald, Lauretta Beaty, Juanita Bolish, Virginia Clair, Dorothy Colburn, Isabel Madison, Rita Moynahan, Dorothy Nichols, Elizabeth Shea, Janice Tate, Vi Thurn, Evelyn Tomlinson, and Rena Wilkes. And nine more graduated on June 1: Grace Birge, Marion Carlstrom, Lillian Conner, Helen Fremd, Virginia Hill, Dorothy Kocher, Virginia Luttrell, Katherine Loft, and Esther Poole.[16]

Sixty-nine women pilots were eligible to ferry pursuits as of June 1, 1944. The April count of thirty-five was virtually doubled.

Chapter 23

The First Transfers

On her way to Orlando for OTS, Nancy Love made a stop in Washington, D.C. She had been asked to give a deposition for the Ramspeck Committee and agreed to do it on Sunday, April 16. Knowing Nancy was due in Washington—and that she was on her way to Orlando afterwards—Jackie Cochran invited Love to be her guest in her D.C. apartment the weekend of the deposition. They had much to discuss.

General Tunner no longer had sufficient trainer-type airplanes for the Class I WASP to ferry. In April, he told Love and Cochran to work out the transfer of non-pursuit-qualified WASP ferry pilots back to the Training Command. Their April meeting marked the beginning of several reassignments of women pilots that would occupy considerable amounts of both Cochran's and Love's time and energy over the next three months.

The Ferrying Division's mission—the ferrying and transport operations within the United States—was absolutely critical to the training of the male pilots who were needed overseas. If the FD's women pilots were not able to ferry pursuit aircraft, they were a detriment, taking on-the-job training slots away from male pilots badly needed overseas.[1] It took General Tunner a while to convince AAF Headquarters that if the Ferrying Division continued to receive women from Sweetwater not ready to begin transition into Class II twin-engine trainers and ferry them cross country, he would have a glut of women pilots stuck ferrying light single-engine trainers. That would severely impede the male pilots' training program. That was why, looking ahead in September 1943, he asked—and finally received authorization in December—to acquire only the top fifteen graduates from each successive class at Sweetwater.[2]

Some members of the Air Staff who were close to Jackie Cochran felt the WASP were not being given an equal opportunity for advancement and questioned the FD limiting its number of WASP.

"General George has repeatedly emphasized that male and female pilots must have equal opportunity to transition and progress... that there will be absolute equality," General Tunner wrote. "Such a policy is now, and always has been in full effect. Many WASP pilots have progressed upward, and no doubt, will continue to do so during the forthcoming months. ...WASP [are] given every chance for promotion, advancement and transitioning to heavier type equipment."[3]

The Training Command—rather than sending fifteen new graduates—at first began to fill those slots with WASP who had been refused earlier by the ferrying bases. Those women had been reassigned to bases where they had accumulated additional flying experience. This practice began with the graduation of 43-7 on November 13, 1943. No new graduates would be given directly to FD/ATC until the new training regimen had had time to kick in. Class 44-2 had entered training on September 6, 1943, and was the first class to complete the revised curriculum. Fifteen 44-2 graduates were sent to the Ferrying Division following graduation on March 11, 1944.

In the meantime, some hard feelings erupted because many of the women sent to the FD in November, December, and February were those who had been rejected earlier. This caused the FD concern because it had felt some of these women simply were not qualified to ferry and never would be. It didn't mean they couldn't fly, but simply that they were not ferry pilot—and ultimately pursuit ferry pilot—material. From the women's perspective, some were put right back at the same base that rejected them and they were most unhappy about that.

When the first classes of future pursuit pilots began training in Palm Springs in December, Nancy Love made this recommendation to General Tunner:[4] "When a pilot is judged by transition to be superior, every effort should be made to give her an instrument course and to make her a thoroughly competent Class III pilot, after which she should be sent to pursuit school." In that same letter, Nancy urged that no WASP be pushed to fly pursuit if she did not want to. The Ferrying Division took steps to improve the training situation at the ferrying bases, again, by confirming that women were allowed to fly as co-pilots with men on deliveries, thus offering C-47 as well as B-25-type transition. Delivery flights could be considered training flights.[5]

Mrs. Love did suggest—following the discussion at the January meeting with her squadron leaders in Cincinnati—that priority in training be given those who *did* want to fly pursuit. She also recommended that WASP willing to fly pursuit, and who were judged capable of being trained to do so, sign a statement expressing their eagerness to go to Pursuit School *before* giving them transition to Class III and an instrument rating. And she added that WASP would not be frozen on single-engine pursuit. After accruing hours and experience in single-engine pursuit, they could progress to Class IV aircraft, which meant flying the A-20, A-30, and P-38—high-performance, twin-engine types. This is exactly what she had relayed to her four squadron commanders at the January meeting in Cincinnati.

"There is a genuine need for WASP in both single- and twin-engine pursuit categories, and those WASP who are capable of flying these types will be performing the greatest service for their country and for the Ferrying Division."[6]

What Love and Cochran were working on the weekend of April 15–16 was the first major change in base assignment within the Ferrying

Division. Because Wilmington and Long Beach no longer had sufficient trainer airplanes to be ferried, WASP who still needed on-the-job transition time were to be transferred to the other ferrying bases. The situation was most acute at Wilmington. Squadron commander Betty Gillies could not keep all her women pilots busy. She needed to send up to forty women to other bases and other jobs, where they could be better utilized and could get more flying time.

At their April 15 meeting, Love and Cochran hammered out the following transfers. A letter from Ferrying Division Headquarters to the payroll department, containing the names and postings, is dated April 26, 1944.[7] Transferred from Wilmington to form and lead the new 33rd Ferrying Group/ ATC, located at North American Aviation, Inc., of Kansas in Kansas City was Helen Richey, former ATA pilot and a 43-5 graduate. [8] With her went two experienced ferry pilots, Magda Tacke from 43-1, and Marie Muccie from 43-2, and seventeen pilots from later classes. These twenty women would ferry BT-15s and serve as copilots ferrying B-25s. Ten more were transferred from Wilmington to Romulus. And there was the new squadron at Palm Springs. Six Wilmington pilots were sent there. Those various moves reduced Betty Gillies's Wilmington squadron by thirty-six pilots.

For Marie Muccie, stationed in Wilmington, the move to Kansas City was a godsend. "'We're going to give you a choice.' That's the word that came from headquarters, spring of 1944. 'What do you want, pursuit or bombers?'

"I knew I didn't want pursuit. I liked the idea of having two engines out there or three or four. And I was amazed when I was accepted to go to Kansas City, Kansas, where the North American Plant made the brand new B-25s."[9]

Long Beach, the other base without trainer-type airplanes to ferry, sent eleven of its women to the new ferrying base in Palm Springs.

Original WAFS Esther Nelson was transferred from Long Beach to Kansas City to assist Helen Richey with squadron leadership. Avanell Pinkley from 43-2 arrived in KC from Dallas on May 1. Pinkley and Richey went on to qualify as first pilots in the B-25.[10]

On April 27, after that transfer was complete, Betty Gillies learned that she would receive seven more women when Class 44-4 graduated May 23. Betty and her commanding officer asked that the seven not be sent, pointing out that thirty-six already had been transferred out of Wilmington. Her request was honored. Those 44-4 graduates went to Romulus and Dallas.

The end of May, the Ferry Command tried to transfer thirty-two more women back, but General Harper, signing for General Arnold (who was in England preparing for D-Day), said that "insufficient time has elapsed to test fully their usefulness to the ATC."[11]

This was counter to Cochran and Love's April 15 agreement.

On June 8, Tunner took the situation in hand. He wrote to Arnold and pointed out that the Ferrying Division had determined that a pilot needed *five hundred hours* before he/she was sent to Pursuit School. He requested that, from then on, the fifteen per month quota from the Training Command be women with five hundred hours.

No graduates were sent to the Ferrying Division after 44-4.

As for Nancy Love's deposition, on Sunday April 16, 1944, she met with a Colonel McCormick and a Mr. Shillito, investigators for the Ramspeck Committee. Her answers were straightforward.

Q: What do you think of the qualifications of the girls who come from the training school?

A: They are good pilots ... on a par with what I know of the male graduates of the Training Command.

Q: What is the policy on pursuit pilots?

A: We need pursuit pilots.

Q: (He) was very interested in the relationship between Miss Cochran and myself—wanted to know if there was a battle between us.

A: Miss Cochran's job is an administrative one and mine is an operational one, and there is no battle between us.

Q: Do the girls want militarization?

A: The majority do.

Q: Why do the girls want militarization?

A: For recognition and protection. A civilian girl going into a modification center to pick up an airplane is open to suspicion as a spy. Also, compensation and insurance for the families would then be available to them.

Q: How much do you think that the uniform is worth?

A: I have no idea. I have never been interested in clothes.

Q: What items are issued and what are purchased?

A: We are issued two winter and two summer uniforms. We purchase accessories, etc.

Q: What is the pay of the other operational WAFS [WASP]?

A: $250 per month, plus overtime, which comes to about $280.

Q: Who are the girls who are competing for the Air Medal?

A: I have no idea what you mean. It has been awarded to Miss Erickson.

Q: How was [Miss Erickson] able to make those deliveries so fast?

A: The weather was good; she is a good pilot and a hard worker; and she had fast airplanes.

The six-page document is dated April 17, 1944, and signed Nancy Harkness Love, Executive for WASP, Ferrying Division, ATC.[12]

From Washington, Nancy went to Orlando for OTS.

While Nancy, Betty Gillies, and several other original WAFS were attending OTS in Orlando in late April / early May 1944, Betty's assistant, Esther Manning Rathsfelder, was getting ready to go to pursuit school at the new facility in Brownsville. She had earned her Class II rating in a UC-78 in January and her Class III rating March 26, flying New Castle AAB's C-60. She was assigned to the class beginning May 1.

From Brownsville, she wrote several letters to a friend back at New Castle describing her experiences with pursuit transition.[13]

> May 3, Well, here I am in the land of cactus and bugs. Hot, hot, hot. Dirt, dirt, dirt. Bugs, bugs, bugs. Will have to admit that I've seen more attractive spots. The girls were certainly justified in griping about this base—it's horrific and practically impossible to get decent food. Only a bed for furniture in the rooms and the linen on that is dirty.
>
> Am sitting out on the flight line at the auxiliary field, a dried-up pasture. Put in an hour in the BC-1 this morning, mostly air work. It's extremely pleasant to have nothing to think about but flying and to be able to spend my leisure time sitting on the flight line. This is the first time I've been able to do that and I think it's going to help. [Esther's job back at New Castle often kept her tied to a desk.]
>
> May 4: Margie Gray (43-1) washed out yesterday. Didn't pass her check in the AT-6 and is now on her way home [to Wilmington].
>
> We don't get any time off, except Sundays—if we're caught up on our flying, in which case we go to the beach. It's 30 miles away. Not that we'd have time during the week anyway since our day doesn't end till 6:30 or 7:30, depending on whether we're on the afternoon flight schedule or the morning one.

Flew for an hour in the BC-1. Had to choke down my fear, especially of the oxygen mask and I have a feeling my instructor guessed it. However, I feel a little better about the flying now and I am praying that I will continue to improve.

May 6: Today was beautiful so far as the weather was concerned —most unusual for down here, clear and cool. Flew an hour and twenty minutes this morning—mostly landings, which were all right but my coordination was still off. Didn't feel any better about the altitude work, and didn't do anything—just climbed up and came down again because the manifold pressure behaved strangely.

May 7: Had today off. Slept until noon.

May 8: Routine day. Flying went a little better I think, I enjoyed it more.

May 9: I'm exhausted from the heat in addition to being as stiff as a board from my first PT in over a year. Ground school all day, and all evening and no flying. And, to top it all, a check ride tomorrow. My instructor says it isn't, but I'm sure a ride with the Flight Commander isn't just to put in extra time as he says. I've got to do a good job, I can't fail here—want to fly the peashooters.

May 10: I don't think I've been doing a very good job. I still harbor my passionate fears of pursuit—aggravated perhaps by the ground school training here, which consists in large part of telling you how to bail out. Not that I'm not glad of that. I never knew before, and am now beginning to regard my chute as something besides a cumbersome piece of equipment to carry around. They totally stress it down here. However, I'm dying to fly the little devils now and will be heartbroken if I don't make it.

May 11: Well, I didn't make it and I couldn't be further down in the dumps. Flew four times yesterday—two check rides—and although Lt. Davis, the check pilot, said I was on the ragged edge

> of qualifying, he couldn't do it. He was swell. Recommended more AT-6 time and said I could come back here by July 1st.

Esther was assigned to go to OTS in Orlando soon after returning from pursuit school.

Chapter 24

21st Ferrying Group, Palm Springs

Palm Springs, California, and the 21st Ferrying Group—in addition to being the site of the FD/ATC's Pursuit School December 1943 through March 1944—also was home to two different contingents of WASP ferry pilots. The first group of sixteen women—six original WAFS and ten WFTD graduates—was stationed there from September 30 to November 17, 1943.[1] Evelyn Sharp was appointed C.O. The women were to transition into the AAF's twin-engine workhorse transport, the C-47—one of the requirements for pursuit school. They were to ferry the C-47 as well as BT-13s. All were sent elsewhere in mid-November 1943 as the official Pursuit School was readied for its December 1 opening.

Then while pursuit classes were in process beginning in December, higher-ups debated whether to move the school. Finally, the decision to use Palm Springs for ferrying won out. Palm Springs graduated its seventh and last pursuit class the first of April 1944. Twenty-nine women had earned their Class "P" distinction as had more than 150 men.[2] Pursuit School was moved to Brownsville, Texas, April 15, 1944. The second squadron of WASP pilots—some pursuit qualified and some pursuit-pilot-aspirants—arrived to become part of the 21st Ferrying Group.

Evelyn Sharp was Nancy Love's choice to command the new WASP ferrying squadron, but, learning of Evelyn's April 3 death, Nancy immediately summoned Byrd Granger and Mary Lou Colbert from Dallas to Cincinnati. The three met April 4 and Nancy appointed Granger squadron commander. Colbert would be the Executive Officer and Vivian Cadman, 43-5, also from Dallas, the Operations Officer. The three women would be transferred immediately from Dallas to Palm Springs.

In March, Granger was injured trying to catch the "ferry-go-round" base transportation at Love Field. Loaded down with her "parachute, briefcase, and small bag," she tried to leap aboard the moving vehicle pulled by a small tractor. "I struck my left hip against a corner stanchion. The immediate pain was intense."[3]

Over the next several days, though in pain, she continued to ferry aircraft. Then Nancy called her to Cincinnati, where the pain continued to worsen. She and Colbert returned to Dallas, where Granger went directly to the hospital. Her fight to recover and rejoin the squadron is told in her book, *On Final Approach.*[4]

Nancy's response was quick. She placed Colbert in charge.

"Viv and I went to Palm Springs together, April 18, to start up the squadron from scratch," Mary Lou recalled. "Byrd was in the hospital and then was laid up a long time. I had to set up everything, get a jeep, organize the office. Being the CO was also why I stayed an extra month or so after deactivation to wrap everything up. Byrd didn't join us until very late."[5]

In mid-April, Mary Trotman (43-2)—a Palm Springs pursuit school graduate—was transferred from Wilmington to the new 21st FG squadron. With her went five other pilots from the 2nd FG, part of the group being moved out of Betty Gillies's squadron in order to give them more flying time. One, Eleanor Alexander (43-6), would graduate pursuit from Brownsville.

Photo 34. Martha Wagenseil, Class 43-2, Palm Springs ferry pilot

Courtesy WASP Archive, Texas Woman's University, Denton, TX.

May 1, Vega Johnson, 43-1, and Marty Wagenseil, Ruth Franckling, and Barbara Russell—all 43-2—graduated in Brownsville's first pursuit school class and were transferred from Long Beach to Palm Springs.[6] Other Long Beach transfers, Margaret Castle, Rebecca Edwards, and Lola Perkins (all 43-6), would qualify in pursuit that summer.

Those eleven women—plus Granger, who joined them for the latter half of the squadron's existence, and July transfer from Dallas, Cappy Vail Bridge—became the WASP squadron of the 21st that served until deactivation. Thirteen—pursuit pilots all!

Martha "Marty" Wagenseil (43-2) was assigned to Long Beach upon graduation from Houston. She made her first delivery on July 5, 1943, with Geri Lamphere (43-1), Lewise Coleman, Iris Cummings, and Zelda Lamar (43-2 classmates).

"We were humiliated when we arrived at Tucson because the Tucson tower, on hearing our voices asking for landing instructions, called out in his male voice, 'Clear the air, there are women pilots coming!' This was only the first of many similar experiences.

"We were assigned to flying two weeks with SNAFU airlines. These were interspersed with BT-13 and L-6 Grasshopper deliveries. The supposed two weeks of flying with SNAFU as copilot stretched into longer weeks and months, but that flying was one of the best things that ever happened to me. The pilots were 'old line pilots,' not military pilots, not neophytes, but men with thousands of hours and years of experience. Many of them were generous with their flying time. One of them, Capt. Bob Donnell who later became an Eastern Airlines pilot, let me make all the take offs, do all the flying, do the navigating, and make all the landings, and while I was flying with him, he checked me out as copilot in various types of equipment, such as C-60, B-25, and B-17."

Delivering BT-13s was the primary responsibility of the women ferry pilots assigned to Long Beach.

"The BT-13 was not a fast airplane, and we were not permitted to fly them at full power because the engines had only 1½ hours or less and we had to break them in carefully," Marty recalled. "The girls were very conscientious about this, so we would go up to Downey in the Red Cross ambulance, pick up our BT-13s, deliver them to Enid [Oklahoma] a few days later, get ourselves to the train station, ride the train up to Tulsa, spend the evening at the closed airport sitting on church pews that were provided, catch the midnight airline that came by. The plane was bound for LA, but sometimes the captain would divert to Long Beach, so we got home earlier. Then it was off to Downey early in the morning to pick up another BT-13 for Enid.

"These were very slow trips, but we didn't mind. We loved it!"[7]

Early in 1944, Marty, Ruth Franckling, Barbara Russell, and Vega Johnson got orders for instrument school in St Joseph, Missouri. "We spent the month of February there—deep snow, cold blowing wind! We lived in a BOQ [Bachelor Officer Quarters] with a pot-bellied stove. At intervals during the night somebody would come in and stoke up the stove. It didn't bother us much because we did most of our flying at night in unheated B-25s, C-47s, or C-49s. We wore fleece-lined leather flying clothes, but the temperature in the airplane was the same as that outside, minus 30 degrees Centigrade.

"Three of us flew at a time with one instructor. While one was in the right seat, the other two would huddle in the back, clinging to each other for the extra body warmth we could get. Vega Johnson was one of my flying partners and we came close to a fatal accident one night during an instrument check ride in a B-25. The check pilot was in the left seat, I was in the right seat, the others were in the back. He cut the right engine; I took immediate action to keep the other engine from over-powering the airplane and causing it to roll. It took both of us, with all the strength we had, to control that airplane and get it flying straight and level again. We lost a great deal of altitude."

On April 1, 1944, Marty, Ruth, Barbara, and Vega were among the fifteen WASP sent to the first pursuit school class held in Brownsville, Texas.

"We had tech reps from every company and they had mock-ups of all the fighters," Marty wrote. "These airplanes had single seats. We would have to know how to fly them when we climbed aboard. All pursuit aircraft were there except the P-38 and P-63. The P-63 later took the place of the P-39. The P-38 came later for me at Palm Springs."

May 1, the four were on their way to Palm Springs, to be part of the new WASP ferrying squadron.

Cappy Vail (43-2) married her fiancé Art Bridge in June 1944. They went to Palm Springs for their honeymoon because Art wanted to check out in a P-51. Cappy, a graduate of one of the earliest Palm Springs pursuit classes, had been flying them since January. The men in Art's squadron could not believe that his wife was flying planes that he had not yet flown—particularly the P-51.

"Art finally got his P-51. For awhile, I had more flying time in it than he did—but not for long."

Cappy ferried out of Dallas from June 1943 to July 1944. "We spent a lot of time picking up the new planes at the modification center in Fort Worth and flying them over to Love Field. That first flight was a test flight and from there the aircraft was re-entered into the Dallas schedule and flown out to another base. Sometimes one of us took it, sometimes someone else did."

But things had changed for Cappy.

"When Art went back overseas, he was escorting the B-17s over Japan, meeting the enemy head on. I felt that I needed a change from Dallas. I hated to leave my friends, but I didn't see them that much when we were out on the road. I could tell I felt differently. There was not that freedom, that carefree type of thing."

Photo 35. Cappy Vail Bridge, Class 43-2, Palm Springs ferry pilot, one of the WASP to ferry the P-38 Lightning.

Photo courtesy of the Iris Cummings Critchell Personal Special Collection

She opted for Palm Springs—the smallest of the WASP squadrons and the answer for a fully qualified pursuit pilot like Cappy looking for a change.

"Palm Springs concentrated on fighters, and I'd had a wonderful time in the fighter school there. The girls who were there were like me. They just concentrated on moving the fighter planes out of Long Beach to wherever they needed to go. It was sort of a sub-base of Long Beach. I checked out in P-38s there, and they were wonderful planes. But the P-51 is still my favorite."[8]

After Viv Cadman went to Palm Springs, she earned her instrument rating there flying a B-26. Pursuit school was next and when she returned from Brownsville, she joined the rest of the pursuit pilots at Palm Springs flying what became "THE" routine trip for them.

"We'd go down to Mines Field in Los Angeles and pick up a new P-51 at the North American factory, fly it to Palm Springs in the afternoon, and the next morning head east for Newark, cross country. After we delivered the P-51, if there were P-63s waiting for us, they'd put us on a night train over to Niagara. We'd pick up our P-63. Sometimes you had a buddy, but we didn't count on it. You were on your own. You got your airplane to Great Falls. Usually there was a bunch of us and there'd be a C-47 ready to haul us all back to Palm Springs—where we would get to do the whole thing all over again the next day."[9]

The women's squadron of Palm Springs may have been small, but it was mighty, as Byrd Granger attested in the Commendation she wrote on October 10, 1944, after returning to a leadership role with the squadron. The squadron, she wrote, devoted itself to aircraft delivery and to transition into those aircraft most urgently needed overseas:[10]

> The record is remarkable. Every member of the squadron is a pursuit pilot. The squadron has record of no student washouts in any phase of transition, including pursuit school. In seven months the squadron has had one accident attributable 100 percent to pilot error, one mishap which was partially pilot error and partially

> field conditions, and three serious mechanical failures in which pilots handled their ships so skillfully that no damage resulted to any of the aircraft involved.

The women dealt with a variety of engine failures, sabotage attempts, structural failures, and fire aboard the aircraft, Granger noted, and credited "hard work, expert flying and unswerving devotion to duty exhibited by every member of the WASP Squadron, 21st Ferrying Group, ATC."

Chapter 25

D-Day, the Sixth of June

The events of the D-Day invasion of Normandy and the mainland of Europe dwarf all other news to come out of June 1944; nevertheless the WASP as a unit had its noteworthy share. For the women of the Ferrying Division, significant was the re-ordering of the TDY contingent sent from the 2nd Ferrying Group in Wilmington to ferry the P-47s built at Republic Aviation, in Farmingdale, Long Island.

Since summer 1943, Betty Gillies, Helen Mary Clark, and Teresa James had teamed up with the male pilots from the 2nd FG to ferry Thunderbolts out of Farmingdale. Beginning in January 1944, the three women were joined by a steadily increasing cadre of women pursuit school graduates stationed in Wilmington. These women began to take P-47s from Farmingdale to the docks at Newark, as well as to the modification facility in Evansville, Indiana, and occasionally to the West Coast.

As of June 8, 1944, WASP had replaced all the male pilots at Farmingdale. The task of ferrying newly built P-47s from the factory to their next destination became a woman-only operation under Betty Gillies's guiding hand. Sometimes she was one of the women on TDY—leaving

Helen Mary Clark and Esther Manning Rathsfelder in charge in Wilmington; sometimes she was back at Wilmington and Helen Mary was in Farmingdale. Each of the women ferry pilots took her turn—two weeks on Long Island, then two weeks back in Wilmington. WASP from Romulus also became part of the rotation.

By June 8, the 2nd FG had fifteen WASP pursuit-qualified pilots: six original WAFS—Nancy Batson, Helen Mary Clark, Betty Gillies, Helen McGilvery, Gertrude Meserve, and Teresa James; six Houston graduates—Virginia Alleman, Emily Hiester, Rita Moynahan, Esther Poole, Jane Straughan, and Ruth Grimm Trees; and three Sweetwater graduates—Dorothy Colburn, plus two transfers from Romulus, Tex Clair and Janet Zuchowski. By early July, Sweetwater graduates Ruth Anderson, Celia Hunter, Jo Pitz, Mary Rosso, and Mary C. Wilson had joined them, for a total of twenty pursuit pilots.

On July 15, the last original WAFS to attend pursuit school, Sis Bernheim, graduated in Brownsville and joined the contingent ferrying out of Farmingdale. And there were more Brownsville graduates still to come.

Betty Gillies, early on, convinced her C.O. in Wilmington to give the WASP contingent at Republic Aviation a C-60 transport in which to fly the ferry pilots from Newark back to Farmingdale. Nancy Batson, Teresa James, and Gertrude Meserve checked out in that aircraft and traded off pilot/copilot duties. The C-60 followed each flight of the P-47s from Farmingdale over to Newark and returned all the pilots safely in the belly of the larger, twin-engine craft so they could repeat the process. In the summer four trips a day were possible, provided the P-47s were ready for delivery and the weather cooperated.

The daily drill was to fly the 17,500-pound pursuit from the factory to the docks at Newark—fifty miles, give or take, as the crow flies. Once the airplanes were delivered to the Newark airport, workmen removed the wings and "pickled" the aircraft to protect them from the salt spray on the ocean crossing.

The flight to Evansville took three-plus hours from Long Island (roughly 900 miles). Cruise speed in the big pursuit is rated at 350 mph, but these were new airplanes and the WASP were told not to push the engine. At Evansville, the pilot would leave that airplane to be modified and pick up a finished one—possibly headed back to Newark, possibly headed to the West Coast, or any place within the forty-eight states. A ferrying flight from Evansville to the West Coast, given traveling with the sun and with favorable weather, took two days. Ferry pilots did not fly at night. From California, the pilot would be given another airplane—probably a P-51—to take back to Newark.

In mid-summer 1944, Nancy Love established a small squadron in Evansville to ferry those pursuits going in several directions from the modification center. She rented a little house so that the women had living quarters near the plant. Fifteen WASP from Wilmington and Romulus staffed it and they rotated in and out, serving a two-weeks-at-a-time TDY.

Enough of the women were ferrying the modified P-47s to Oakland, California, for shipment to the Pacific that a SNAFU C-47 would pick them up and fly them back overnight to Evansville. The C-47 had aluminum bucket seats that, in the cold air aloft, made for an uncomfortable journey back and little sleep. Margaret Ann Hamilton, one of the Romulus contingent who did TDY at the mod center, devised her own system of keeping warm. She learned to squeeze her slight frame into her parachute bag—knees against her chest—then had another WASP zip her in.[1]

Initially, Betty Gillies placed Nancy Batson in command at Evansville, but Nancy hated the job. "I just wanted to fly. I didn't want to be sitting behind a desk and in command of anything. I finally got Betty to transfer me back to Wilmington and up to Farmingdale, but I think she decided to punish me because for awhile all I did was deliver dogs—war-wearies—to the bone yard." Helen Mary Clark then took over the command of the Evansville squadron.[2]

The second of Nancy Love's pursuit ferry pilots died June 11, 1944.

Jary Johnson from Romulus and Dottie Nichols from Long Beach—the two friends from California who had gone to Houston together as members of Class 43-2—were in Bismarck, North Dakota, en route from Niagara Falls where they had picked up P-39s to be delivered to Great Falls, Montana. The ceiling was low that Sunday morning. All the pursuits were grounded, waiting for the fog to burn off. Several WASP and male ferry pilots were on that trip. Since it was Sunday, Dottie decided to go to church, and she asked Jary if she wanted to go. Jary, surprised, said "yes." She had never known Dottie to be a churchgoer.

Back at the airport, when the ceiling finally lifted, the pilots reassembled, checked their aircraft and flight plans, climbed in, and did their cockpit checks. They were ready to go and one by one took the runway.

"Dottie's engine quit on takeoff," Jary recalled. "Right over the runway, about twenty to twenty-five feet in the air, which is a very critical time to lose an engine. She didn't have enough flying speed. She went nose down, straight into the runway. It burned up immediately"[3]

A wingtip fuel tank burst into flame on impact. The control stick pierced Nichols's forehead. She was dead before the flames touched her.[4]

"Because I was her friend, they had me accompany her home," Jary recalled. "Strangely enough, I just happened to be there and I saw it happen. That was a total coincidence, because we were based at two opposite ends of the country and usually flying different planes on different routes. We just happened to overlap on that route and we had been rooming together in a hotel in Bismarck at the time she was killed.

"I had to deliver her body by train to Los Angeles, to her mother, which was difficult. They sent a man from Romulus to pick up my airplane after I left it in Bismarck. He was to deliver it to Great Falls. The engine quit, but he wasn't killed. He managed to land it in a field and it damaged the airplane, but there was no fire and he didn't get hurt."[5]

Six WASP from Long Beach, Dottie's home base, served as pallbearers for the interment at Forest Lawn. Both Nancy Love and Jacqueline

Cochran were in attendance.[6] Dottie Nichols had graduated from pursuit school in Brownsville four weeks earlier, on May 15.[7]

Mary Catharine "Jary" Johnson graduated from Palm Springs pursuit school February 21, 1944. She was one of the earliest to complete the course, but there had been that awful moment when she thought she had blown it.

"After I put that Stinson in that New Jersey field, it was very iffy whether or not I would get to go on to pursuit school, even though I had been assigned," she recalled. "When you do things like that, they are apt to take you off the better planes. I went through agony. Back at Romulus, they sent me to Colonel D'Arce's office.[8] I was shaking in my boots.

"He said, 'well, in view of your previous performance, we're going to send you to pursuit school regardless.'

"So I've always thought D'Arce was wonderful."

By June 11, 1944, Jary had numerous pursuit deliveries in her logbook. "I was put through instrument school, and then sent to Palm Springs to learn pursuit flying. There were four women in my class at Palm Springs: Jane Straughan from the first class; Florene Miller, a WAFS, and Emily Hiester and me from 43-2. Our class picture showed forty-six men in uniform with four women sitting in the middle of the front row.

"After graduation, they took us to Long Beach and we were each assigned a P-51. That was our way of getting home—to fly P-51s to Newark. So my very first ferry trip after pursuit school was in a P-51, and I went all the way across the country.

"Everybody likes the P-51. They're so smooth. Stationed in Michigan, I didn't get to fly very many. Our shuttle run was from Niagara Falls to Great Falls, Montana in P-39s and later P-63s

"I ferried all the single-engine pursuits, because, when we delivered an airplane, sometimes we were sent back to our base and sometimes to another factory. Frequently, I was out for six or seven weeks at one time without ever getting back to the base. Every time I delivered an airplane they'd take me someplace to pick up another one. I was just ferrying from here to there to there. I got to ferry some P-47s out of Evansville, Indiana, and some P-40s from the Curtiss plant in Buffalo down to Florida, and of course the Bell planes, the P-63 and the P-39.

"I didn't get to fly the P-38. I'm kind of sorry I didn't, but it was not available at Romulus."[9]

D-Day—June 6—launched the Allied offensive that, over the next eleven months, would make its way across western Europe to Berlin, bringing an end to the war in Europe. The day before—June 5—the WASP's militarization troubles came to a head. That was the day the Ramspeck Report was released. Very briefly, and paraphrased, it said:[10]

- The WASP program was "costly and unnecessary." It recruited inexperienced young women.
- It made possible the "dismissing of / failing to utilize" male instructors.
- It lowered standards for women while keeping them high for men.
- The program was experimental.
- Using women to alleviate the "claimed" manpower shortage was not working.

The committee recommended that Congress shut down the flight training program but—and this was critical—retain WASP already actively serving, and those in training be allowed to complete the course.

On June 21, the House Committee on Military Affairs sent HB 4269 —the WASP bill—to the House of Representatives for debate. The bill was doomed from the start. Rep. Joseph O'Hara of Minnesota called it "social legislation to glamorize the war."[11]

Congressman Costello, the sponsor of the bill and spokesman for the War Department, gave the following very simple explanation of the bill.[12]

> The sole purpose of this bill is simply this, to take these women who are now with the Army Air Forces in a civilian capacity and convert them into a military capacity. That is the sole purpose of the WASP bill, and nothing else.

The bill died on the House floor.

Jacqueline Cochran, in her report to General Arnold in August, pointed out that, by the time the WASP bill was defeated, Congress had already passed the Army Appropriations Bill, which included slightly over $6,000,000 in funds for the WASP program.[13]

The Ramspeck Committee did not object to using women pilots. What it objected to was the recruiting of inexperienced personnel.[14] The committee was on record as recommending that qualified WASP would continue to serve and their salaries would be adjusted in accordance with their experience and responsibilities. Insurance and hospitalization would be made available to them. These were the very reasons the WASP had sought militarization in the first place. All this, of course, went for naught.[15]

No new WASP training classes would begin at Avenger Field. By June 30, forty-eight already had reported for Class 45-1. Hasty arrangements were made for the Domestic Transportation Division of the ATC to return the girls to their homes. The five classes at Sweetwater were allowed to finish, graduate, and be assigned to active duty.[16]

A pivotal month, June 1944.

Chapter 26

A Typical Coast-to-Coast P-51 Delivery

Where you were stationed as a Ferry Command pursuit pilot determined what aircraft you flew much of the time and also where you delivered those aircraft. The women pursuit pilots of the 2nd Ferrying Group in Wilmington spent the majority of their time in Farmingdale, New York, and ferried Republic Aircraft's P-47s to Newark. WAFS Teresa James described it as "The 'gear up, gear down' flight. Thirty minutes in the air, max, unless the fog came in. Then it was circle, circle, circle. There was always smoke and haze over the New York City area."

Romulus ferry pilots, the 3rd Ferrying Group, spent most of their active duty time ferrying first P-39s and later P-63s from the Bell Aircraft factory in Buffalo/Niagara, New York, to Great Falls, Montana. Those aircraft were destined for the Russians via Alaska, but the women only took them as far as Great Falls. "They wouldn't let us fly to Alaska," Jary Johnson, 43-2, Romulus pilot said.[1]

Dallas pursuit pilots, the 5th FG, took P-51s, built in nearby Grand Prairie, Texas, to Newark, and P-38s modified at Love Field wherever they needed to go.

The longest cross-country trips were the ones the Long Beach, 6th FG, WASP made from Mines Field (now LAX) all the way to Newark, New Jersey. Those P-51s were built at the North American plant in Inglewood, California. The WASP based with the 21st FG in Palm Springs made those P-51 flights as well. Long Beach pilot Iris Critchell relates the story of one of those flights in July 1944.[2]

"Thirteen P-51s ready to go. Let's get 'em out of here! Palm Springs before sunset." The voice of the Operations Officer immediately brought to attention the thirteen 6th FG pilots—male and female—waiting in the Ready Room.

By mid-July 1944, the ferry pilots of the 6th FG, knew the drill in their sleep. Gin Rummy hands folded, letters home quickly stored, reading material abandoned, those napping were wide-awake. They grabbed their parachute bags and briefcases and headed for the door.

They were, as the room's name implied, ready—ready to climb into thirteen new P-51s and begin their trip east to deliver them.

"A production test pilot flew the first hour check of the airplane," Iris Critchell related, "then they finished that flight by moving the airplane from the North American factory down to Long Beach. They parked them out on the big grassy areas between the runways. Long Beach was always covered with airplanes awaiting delivery."

The ferry pilots' orders: fly those brand-new P-51s east from Long Beach cross-country to Newark, New Jersey, where they would be loaded onto a Liberty ship waiting at the docks. The ship was due to sail for England as soon as its precious cargo was safely on board. The sooner the ferry pilots got the aircraft to Newark, the sooner those P-51s would be on their way across the North Atlantic and into battle.

For months, this scene had been repeated daily. Since January, the women pilots had joined the men in increasing numbers. The sleek, fast, single-seat, single-engine fighters were the secret to the strategic bombing of Berlin and other German cities that housed the Nazi war factories.

"The early 1944 ferrying of P-51s was stepped up when the C and D models came out," says Critchell. "The new C & D models were supercharged internally. They were equipped with auxiliary fuel tanks so they could make the trip to Berlin and back protecting the bombers from attacks by enemy fighters." Their new high-altitude performance was designed to turn the tide of those extremely high losses experienced by the B-17s and B-24s of the Eighth Air Force in England and the Fifteenth Air Force in Italy. Before the long-range P-51s arrived, the bombers suffered terrible losses in their daylight raids.

The ferry pilots from the 6th picked up P-51s daily. The only drawback was the insidious morning fog on the southern California coast. The resulting low ceiling frequently delayed departures until it moved offshore in the afternoon. But when the fog cleared, the ferry pilots could take off and fly the hundred miles inland through the San Gorgonio Pass[3] to Palm Springs, landing before dark. They would RON there.

If Palm Springs was clear in the mornings, they could depart for the East Coast the next day within half an hour of sunrise—as per Ferrying Division regulations.

Such was the daily life of a pursuit ferry pilot stationed with the 6th Ferrying Group in 1944. The largest of the Ferrying Groups, the 6th boasted 1,700 male pilots—seven squadrons—and by then the WASP ferry pilots, the eighth squadron, numbered between thirty and fifty. B.J. Erickson was WASP commanding officer (C.O.) at Long Beach. Ruth Thompson was her executive officer and second-in-command. Iris Cummings served as third in command when needed.

"We had at least four ferry routes we could fly eastbound, depending on weather," says Critchell. "The southern route was Tucson, El Paso,

Midland, Dallas, Monroe, Atlanta, and up the coast to Newark. Frequently we used the route through Albuquerque, Amarillo, Tulsa, Nashville to Newark. The northern route was Salt Lake City, Denver, Omaha, Columbus and on into Newark. That route was used more on delivery flights flown from Oakland or from the Midwest [example, Evansville], usually with two overnight stops. Each ferry pilot was responsible for her own plane, so we all were free to choose our preferred route and altitude." [4]

Because of the lack of oxygen as their flight paths climbed to higher and higher altitudes, Army pilots were required to wear oxygen masks whenever they flew higher than 10,000 feet. To fly these routes in the P 51s and all the other high-performance, high-flying airplanes, WASP pilots also were issued oxygen masks. This meant that they had to go through testing in the high-altitude chamber and be fitted with their own oxygen mask—which subsequently was checked every six months.

Flying east out of Palm Springs on the Salt Lake City/Denver route, they encountered the 11,000- to 12,000-foot-high passes through the Rocky Mountains. Flying those high-speed, high-performance airplanes, the ferry pilots often flew cross-country 20,000 feet or higher.

"Another reason the P-51 pilots wore oxygen masks was to prevent carbon monoxide poisoning," Critchell points out. "Because of the summer heat over the desert, some of the fellows had flown with their canopies partially open. The canopy can be pushed back about one inch to a latch. But in the P-51, the exhaust collector stack on the left side of the engine fed right back into the cockpit. Three of the men went down. Obviously carbon monoxide. They just passed out and went right straight into the ground. So orders were that everyone was to wear oxygen masks and keep that canopy closed. Go up high. So we did. Besides, it was summer in the desert and at 20,000 feet, the air was cooler."

As the number of pursuit aircraft built rose throughout 1944, the Ferrying Division realized it had to meet the ever-increasing demand for stocking the 130-octane fuel required to fly the high-performance airplanes. The solution: set up several key locations for the purpose of maintaining fueling service and minor maintenance for the ferry pilots. Headquarters sent four or five enlisted personnel to staff each of these locations across the country. In the desert, some refueling stops were mere waypoints set up for that very purpose. They had a barracks, complete with an overnight facility if a ferry pilot needed to RON.

Critchell recalls having to spend several nights at the desert refueling stop at Coolidge, Arizona, which is fifty-six miles southeast of Phoenix, and sixty-nine miles northwest of Tucson. "There was no shade. Daytime temperatures can climb to a hundred and twenty degrees Fahrenheit."

After landing for the day, each pilot had to see that proper all-night guards were provided if the airplane carried confidential-level classified equipment. Then, the pilots would head for the Operations Room to check in and send their RONs—"remaining over night" telegrams. Those required the pilot's name and aircraft serial number, the code for their location, reason for the RON, which could simply be sundown, but, if weathered in or down with a mechanical problem, there was a code for that. Then, if they were in a city of any size, they were free to find lodging and have dinner. But if they were at one of the desert outposts, they had to make do with a bunk in the barracks and meals from a field kitchen.

"Landing at Newark, our wheels had barely stopped rolling when the ground crew would grab hold of the tailwheel strut and start moving the airplane toward the boat—sometimes while we were still in the cockpit," Critchell recalled. "We'd have to scramble and get our stuff out. We would watch the airplanes being towed onto the decks of the Liberty ships, because they were docked at the edge of the ramp at Newark. We completed our paperwork stating that the airplanes were accepted as delivered, and then they were no longer our responsibility."

Earlier in 1944, after the WASP delivered to Newark, they might receive orders to go to Buffalo or Niagara where they would pick up a P-40 or P-39. But about three months after starting to ferry P-51s, the D and then the H models were out and the priority on them was tops. "Then we were no longer allowed to pick up another plane on the east coast," Critchell said. "Our orders read: return to base immediately by fastest available transportation.

"My records, orders and logbook show that, by the middle of July 1944, we were very busy shuttling P-51s to Newark and then riding home to Long Beach on the airliners or occasionally on ATC's military transports.

"The priority on the P-51s was very high because of losses. Liberty ships carrying them across the Atlantic were being sunk. So we were put on orders to return to our duty station, Long Beach, immediately, so we could bring more P-51s east to Newark."

LaGuardia Airport was the departure point for American Airlines, which had the routes from New York to Burbank, the commercial airport closest to Long Beach. The women of the New York City Red Cross had developed a volunteer transportation service that met the ferry pilots at Newark and drove them in vans or recons (modified vehicles) from Newark, through the tunnel under the Hudson River, through lower Manhattan and out to LaGuardia north of Queens. Forty, fifty, maybe even sixty ferry pilots, mostly male but some female, came in any afternoon, delivering from points west or south like Atlanta or Columbus.

"The deliveries at Newark typically were toward the end of the day and the tower had its hands full getting all those incoming airplanes safely on the ground. That also meant quite a few ground vehicles were waiting to get all of us over to LaGuardia—there might be traffic jams on the ground and in the air.

"At 10 o'clock at night, I'd be dragging my parachute bag—which also contained my oxygen mask and my personal stuff—across the marble floor of the terminal at LaGuardia, headed for the American Airlines ticket counter. The ticket agents would groan when they saw us coming, because they knew they would have to offload passengers from those airliners."

Ferry pilots traveled on government transportation requests—TRs. They held War-B-II priority, which meant they had priority over all passengers except for the President and his cabinet members. This was true throughout the war because delivery of aircraft was so vital to the war effort. Getting those ferry pilots back to fly another P-51 was critical. Many high-ranking officers, as well as civilians, were bumped because ferry pilots had to get back to base.

Pursuit pilot Vi Cowden (43-4), who was stationed with the 5th Ferrying Group at Dallas Love Field, told of bumping popular crooner Frank Sinatra. When the airliner that was supposed to be carrying him stopped at Memphis, a crowd of adoring, screaming teenage girls waited for him to step off the plane where they could see him. But he wasn't on the plane and they were greatly disappointed. A diminutive woman P-51 pilot dressed in her WASP flying clothes had commandeered his seat.[5]

"Typically they would put us on the 10 PM or the 1 AM or 3 AM flight," Critchell recalls. "The only airliners of the day were the DC-3s that seated only twenty-one people, so we could fill up two or three planes."

The ferry pilots found a balcony over the main lobby at La Guardia where there were some hardwood benches. They usually had three or four hours before the flight departed, so they would stretch out, put their feet on their briefcases and use their parachute packs for pillows. The accommodations were far from comfortable, but ferry pilots learned to sleep wherever they could.

Flying aircraft that carried confidential or secret equipment, the women as well as the men had to carry a 45-mm automatic handgun for security. Many of the P-51s qualified for that confidential category because of the secret IFF radio equipment on board.[6]

"There was a red button on the back of the radio frequency-select button panel which was the 'destruct' button. We were instructed that if we ever had to leave the airplane under less than secure circumstances, or had a crash of some sort, we were to be sure to push that red button to destroy the radio installation," Critchell explained.

"And, yes, to comply with orders, I carried a .45 automatic with me—but not loaded! We were issued two bullet clips, so we very carefully put the gun in our leather briefcases and stowed the bullet clips in another bag so that we at least reduced the possibility of having someone overcome us and use the weapon on us. And since we carried the weapon in our briefcases, we had to sleep with them as well."

By fall 1944, some planes carried early radar identification—also top secret.

Nancy Batson, an original WAFS who, like Iris, flew all the pursuit planes, loved to embellish her gun-toting story. She would tell those who asked her for stories about ferrying planes during the war that their orders were, "In case of capture, shoot the airplane. They weren't worried about us. Those airplanes had secret installations on them—bombsites, radios, radar equipment and the like. The powers-that-be didn't want any of those secrets captured, so they told us to shoot the airplane. Aim for this little red button."[7]

Nancy, an Alabama sportswoman who had done some shooting and was a pretty good markswoman, strapped her sidearm on when she ferried aircraft with secret equipment rather than carry her weapon in her briefcase. A photo of her standing in front of a P-38, another aircraft that carried sensitive installations, shows the holster on her right hip.

The police officers at LaGuardia got to know the ferry pilots and would wake them up in time to catch their flights. The men from the ticket counter also kept track of them.

On the flights back to the West Coast, they flew all night. The first stop was in Memphis or Dallas—where the Love Field ferry pilots got off—but the pilots bound for the West Coast didn't stir until El Paso. The typical wake-up call was a pillow thrown by someone across the aisle. The stewardess would come with breakfast before they landed at

El Paso—powdered eggs and powdered milk, which none of them could stomach after the war.

"Given the time change from El Paso to Burbank, we'd land well before noon," Iris says. "They'd send the recon trucks from Long Beach to Burbank to pick us up and take us back to Long Beach. We staggered into headquarters dragging ourselves and our dirty clothes, then reported back to our squadron and our Ready Room in our barracks. The officer of the day would greet us with, 'Hi, how was the trip?' She'd listen to our tales and then say, 'well, go upstairs and clean up. You're on orders. Come downstairs in an hour and pick them up.'

"So we go up and shower and wash our hair, wash our dirty clothes, pack clean ones, and about three in the afternoon pick up the orders for the next flight. We were beat, and wanted nothing more than a good meal and to sleep in our own cots that night, but the orders were always the same. 'Get out of here before sunset' and it began all over again. We would take the P-51s over to Palm Springs to avoid the morning fog and then we could be on our way to Newark the following morning."

Photo 36. Jill McCormick, Class 43-5, a Wilmington ferry pilot, standing on the wing of a P-51D.

Courtesy WASP Archive, Texas Woman's University, Denton, TX.

Part VII

Winning the War, Losing the Battle

August–November 1944

The WASP of the Ferrying Division not only continued to fly, from June on their number of deliveries soared as progressively more of them were eligible to fly pursuit. Women who could not qualify for pursuit were returned to the Training Command for assignment elsewhere. But the future for the WASP did not look promising.

On July 20, a group of German officers, carrying out a conspiracy, attempted to kill Adolph Hitler. The attempt failed.[1]

July 31, the advancing Russian Army was reported to be twelve miles from Warsaw, Poland,[2] and U.S. troops led the breakout from Normandy —fifty-five days after the D-day landings.[3]

August 4, an RAF flying officer piloting a Gloster Meteor jet fighter engaged a German flying VI bomb in combat—the first time a jet had been in combat with another jet-propelled machine. The fighter's guns jammed, but the pilot overtook the missile, "tipped it over by inserting his wingtip under its wing, and sent it out of control earthwards."[4]

August 24, 1944, Allied troops retook Paris.

Chapter 27

The Great Transfer

The Great Transfer of August 1944 constituted the most significant movement of WASP personnel accomplished during the twenty-eight months of the women's program. The women were never told WHY this was necessary. Rumor, of course, took the upper hand and hard feelings, resentment, and blame had a heyday.

Betty Gillies no longer had trainer airplanes in Wilmington for women to ferry. Her biggest challenge was moving P-47s out of Farmingdale—the work now being done by her all-woman squadron on TDY there. She had twenty pursuit pilots by July 1 and needed more. By October she would have thirty-four. Nancy Love had opened the WASP ferrying unit at the P-47 modification center in Evansville, Indiana, using pursuit pilots on TDY out of both Wilmington and Romulus. That, too, fell under Betty's administration.

Class II pilots had to be close to earning Class III status and a slot at pursuit school. Otherwise, Betty could not keep them busy. Instead of having those women languishing around base attending ground school, not flying, and becoming increasingly disgruntled, Betty wanted them sent where they could fly. That meant going back to the Training

Command where they could be assigned to do a variety of jobs. These included administrative flights (flying non-flying personnel), serving as test pilots on repaired aircraft, and serving as utility pilots flying BTs, ATs, and small twin-engine aircraft. Some also flew copilot in bombers and transports.

Long Beach had the same problem. The squadron's initial assignment had been the pickup and delivery of single-engine BT-13 trainers built at the Vultee factory in nearby Downey. By early 1944, there were virtually no trainers being ferried by the 6th Ferrying Group, but there were pursuit aircraft and larger, heavier twin-engine aircraft in abundance. By early July, Long Beach had twenty-two pursuit-qualified pilots. The squadron topped out in August at twenty-seven.

In Romulus, the pursuit-qualified women were transitioning into the Bell P-63 Kingcobra—a bigger, faster, longer-range version of the P-39 Airacobra. Romulus also provided pursuit pilots for the P-47 modification center in Evansville. Some pulled duty in Farmingdale. Non-pursuit-qualified Romulus WASP flew copilot in the B-24s built next-door at the Ford Willow Run plant. Romulus had fifteen pursuit-qualified pilots by early July. That number grew to twenty-nine by October.

Dallas still moved twin-engine Beeches and Cessnas built in Wichita, but primarily the WASP had P-51s to ferry from North American in Grand Prairie and P-38s to move from the Lockheed modification center at Love Field.[1] Dallas had twenty-two pursuit-qualified pilots by early July. That number grew to twenty-eight.

The 21st Ferrying Group at Palm Springs, as of July 1, had eight pursuit-qualified women pilots. Ultimately, it would have thirteen.[2]

Those figures add up to 131, but the Ferry Command experienced the deaths of four pursuit pilots in 1944, the resignations of several, and some groundings for health reasons, thereby bringing the number actively flying pursuit at any one time down under 120.

Nancy Love gave the women every opportunity—in fact urged them—to volunteer for pursuit school, but she would not "send" a woman who did not want to go. Flying pursuit aircraft involved more risk than the slower, less complicated trainer aircraft. To handle pursuit took skill and an aptitude far and above that needed to fly a single-engine trainer or small twin-engine aircraft. The men who flew pursuit were looked on as the elite of the World War II flying corps. The women who flew pursuit proved to have the same rare qualities—"the right stuff."

Both men and women washed out of pursuit school. The difference was, a WASP who washed out could choose to go back to the Training Command and fly non-pursuit aircraft or she could resign. A male pilot who washed out could not resign. He was sent elsewhere to train for another job.

Early in June, Nancy went to the Air Transport Command's 4th OTU (pursuit school) in Brownsville to attend the graduation of the largest pursuit class yet—nine women and eighty-seven men for a total of ninety-six new pursuit ferry pilots. She told a reporter there that she was pleased with the progress the WASP assigned to the school in Brownsville were making.[3]

On July 3, 1944, Jackie Cochran placed a key phone call to Nancy Love. A transcript was kept of the conversation.[4] The two leaders were, at that point, trying to stop the Training Command from sending several Avenger graduates from Class 44-5 to Ferrying Division squadrons because they would have to be reassigned immediately and sent elsewhere. "Do you have all the girls you'll ever want, or do you have too many now?" Cochran asked.

"We can use more if they're qualified," Nancy answered. "You remember we suggested that the fifteen a month we get be qualified [for rapid transition] to be sent to pursuit school. That part's all right with us. Twelve just graduated [members of the Brownsville June 30 graduating class], so we've got eighty-one pursuit pilots and we have ten going into this next class."[5]

Following the conversation between Love and Cochran, General George, ATC commanding general, received a memo—dated July 17, 1944—from General Robert W. Harper, Air Staff: "A total of 126 WASP are to be transferred from the Air Transport Command to the Flying Training Command."[6]

Immediately after her July 3 phone conversation with Cochran, Nancy Love lost two more of her ferry pilots—both out of Fairfax. Susan Clarke (44-2), one of the Wilmington WASP stationed with the 33rd Ferrying Group in Kansas City, died when the BT-13 she was ferrying crashed near Columbia, South Carolina, July 4, 1944. Paula Loop (43-2), from Romulus and subsequently sent to Fairfax, died when the BT-13 she was ferrying crashed thirty miles from Medford, Oregon, on July 7, 1944.[7]

Nancy Love did receive some good news. On July 17, 1944, Major Kenneth Stiles, Air Corps, Statistical Control Officer sent this memo.[8]

Subject: WASP Deliveries

I believe you might be interested in the fact that the WASP established a new record for completed domestic movements with a total of 953 for June 1944. Of these, 529 were pursuit deliveries.

On August 9, 1944, this memo—WASP Information for Weekly Summary of Activities—was circulated.[9]

Two WASP from the 6th Ferrying Group, Miss Helen Richards and Miss Florene A. Miller, put in 70:50 and 68:40 [hours] respectively on pursuit deliveries [P-47, P-39 and P-51] during the month of June.

On August 15, 1944, the Ferrying Division released 126 (some sources say 123) non-pursuit-qualified women pilots to the Training Command for reassignment. The WASP in the Ferrying Division roster then stood at 140 and that number stayed relatively constant through November.

By the end of September, the WASP were ferrying three-fifths of all the pursuit aircraft delivered.[10] Some forty WASP entered pursuit school between mid-July and mid-September, of whom thirty-one graduated. The last two women to graduate from pursuit school did so October 15, 1944.[11]

"The transfer the summer of 1944 was very simple," says B.J. Erickson London, WASP squadron leader, 6th Ferrying Group. "The girls who had not qualified in pursuits were sent to the Training Command. There were no other airplanes left in Long Beach for them to fly. By sending them to other assignments, they got to fly airplanes they were qualified to fly, doing other jobs at other bases."[12]

"General Tunner made sure that the Ferrying Division requirements applied to both the male and female pilots," says Iris Critchell. "Though orders to transfer were hard for some to accept, I did hear from some of the girls who were transferred out of the Ferry Command. They were flying at repair depots or carrying personnel and goods in small twins for some of the bases. They simply went on and flew what they were asked to fly."[13]

Women involved in the "Great Transfer" included those who had no desire to ferry pursuit; had been eliminated from pursuit school; had failed in transitioning to higher classes of airplanes; were not considered pursuit material; had insufficient experience or did not have time to build the experience before the program was shut down. Two of the original WAFS were included in this group.

If a woman did not want to go back to the Training Command, she could resign. Because the WASP were not militarized, they were free to leave any time and several did. Over the twenty-eight months of the WASP existence, 150 (13.62 percent) did resign.[14] A few non-pursuit-qualified women stayed in the Ferrying Division by moving into administrative positions and a few continued to ferry Class III twin-engine cargo and transport planes, and the occasional AT-6. By then a new category of deliveries had been added to the Ferrying Division's repertoire—war-weary airplanes that had to be ferried to the bone yard or to training grounds to be used for bombing and target practice. This included everything from beat-up trainers to shot-up pursuits and bombers. These were often hazardous ferries and required extra care and expertise in handling. (This is discussed in Chapter 29.)

Fourteen Original WAFS were ferrying pursuit at the time of the Great Transfer. Five original WAFS were non-pursuit pilots.

Lenore McElroy, a Class III pilot with 4,000-hours, was the mother of three teenagers and had opted not to fly the high-risk pursuits. Also, she was Barbara Donahue's second-in-command at Romulus and virtually indispensable. Originals Bernice Batten and Betsy Ferguson were transferred to the Training Command. Nancy Love said of Batten that she was an excellent pilot on smaller ships and highly recommended, but she was so small (a little over five feet and small of stature) that she was unable to handle the heavier, more powerful aircraft.[15] Bernice, an instrument-rated Class II pilot, was transferred to Merced, California, to fly BT-13s and AT-6s, but before the end of the WASP program, she opted to leave and joined the Women Marines. Betsy was sent to Alamogordo AAB and then to Walker AAB in Victoria, Kansas. At both bases she flew copilot in B-17s and did utility and administrative flights in twin-engine aircraft.

When Fairfax was closed in October 1944, Original Esther Nelson and five other Class II and Class III pilots were sent back either to Dallas or Wilmington to ferry everything but pursuit. Pursuit-qualified Helen Richey, the WASP squadron leader at Fairfax, also returned to Wilmington to ferry pursuit out of Farmingdale. And Esther Manning Rathsfelder, a Class III pilot, continued to serve as Betty Gillies's operations officer in Wilmington until her resignation as of October 1.[16]

Betty had lost her second-in-command Helen Mary Clark to the squadron leader job in Evansville in September, so with Esther's resignation, Marjorie Gray (43-1) became Betty's new second-in-command. Marge did not fly pursuit, but flew nineteen other aircraft types. She was indispensable to the WASP squadron; therefore she remained with the Ferrying Division.[17] Betty designated pursuit pilot Jane Straughan

(43-1) WASP Executive for Operations and Training in a memorandum dated October 6, 1944.[18]

Brownsville would graduate thirty-one more WASP pursuit pilots between July 15 and October 15. No more women were sent to pursuit school after mid-September. Added to the eighty-nine eligible to ferry pursuit as of early July, this gave the Ferrying Division 120 WASP pilots qualified to ferry pursuit aircraft.

The statistics available for November 1944—the last full month during which the WASP were ferrying—lists 140 women on the duty rolls. Their class qualifications are: Class V—four; Class IV—29; Class III—102; Class II—four; Class I—one. A total of 113 of the 140 were flying pursuit.[19] (Another source says 117.[20])

Four Ferry Command women reached Class V and qualified as first pilots on four-engine bombers and transport: Original WAFS, Nancy Love, Betty Gillies, Barbara Erickson, and Delphine Bohn. All four also were pursuit —“P”— qualified.[21] Of the twenty-nine holding Class IV ratings, those who ferried the P-38 carried the “P” designation, so they were Class IV-P. The remainder of the Class IV pilots were qualified on the B-25 or A-20, but did not fly the P-38.[22]

Ability and the time to work one’s way up to qualify were the keys to pursuit transition. Some of the women simply did not have sufficient time to qualify.

Of the 126 (the count this author gets is 123) ferry pilots sent back to the Training Command in August 1944, two were original WAFS; from Class 43-1, one; 43-2, two; 43-3, three; 43-4, twenty-seven; 43-5, twenty-three; 43-6, twenty-three; 43-7, eight; 44-2, six; 44-3, fourteen; 44-4, thirteen.[23] Classes 43-4, -5, and -6 accounted for an excessive number—73—but with good reason. Those three classes sent by far the greatest number of women to the Ferry Command—214, by the author’s count.

Two women who were *not* sent back to the Training Command for reassignment were Ruth Adams from 44-2 and Maurine “Mimi” Miller from 44-4—the only women from the 1944 graduating classes to be sent to, and graduate from, pursuit school.

Photo 37. Ruth Adams, Class 44-2, Wilmington ferry pilot.

Courtesy WASP Archive, Texas Woman's University, Denton, TX.

Ruth thought flying was dangerous but saw a sign at a small airport offering 15 minutes of flying for $3.50. "For about five miles I mumbled to myself about how stupid this was. But then I asked myself, 'Are you afraid?' I turned around and went back." The "intellectual challenge" led her to earn her private license through CPT at Northwestern University.[24]

"Flying gave me a lot of courage." Still, Ruth admits that nerves and lack of self-confidence periodically got in her way as she went through training. But if she screwed up, she fixed it the next time she went up. Ruth swore by her checklists, which she says got her through pilot training. Ruth had a photographic memory. This made reading the instrument panel—understanding what all those dials and gauges were telling her and remembering it—quite easy for her.

In primary training, she was told to "keep her head on a swivel"—the term used at Sweetwater for watching for traffic. She followed that instruction to the letter and failed her first check ride. She moved her head so much she made her examiner nervous. Her primary instructor, Ray Booth, stood up for her. "She's a good flyer," he insisted. Ruth got a second check ride. The same examiner took her up again. This time, when they got down, he said, "She's a honey." She passed.

Ruth and fourteen other 44-2 graduates were sent to the Ferry Command. She was assigned to Wilmington, but quickly dispatched to the 33rd Ferrying Group at Fairfax Field in Kansas City. Working with Helen Richey, legendary pioneer flyer from the 1930s who was the WASP C.O. there, had a profound effect on Ruth. "Helen talked a lot about flying. She told us, 'You do whatever you have to do to make the plane do what you want it to.' And I always remembered that!

"In June we heard they wanted volunteers for Brownsville. For personal reasons, I wanted to be stationed back in Wilmington. I figured the way to get there was to go to fighter school because the gals at Wilmington were flying pursuit."

Ruth volunteered and was sent to Brownsville in mid-July 1944. "You never knew what you would be assigned to when you went to the flight line. We never had an instructor. You read the manual. I sat in the planes on the line and went through the checkout procedures so I'd know where everything was. The location of the controls was slightly different in each airplane. The manual told what the settings should be—the manifold pressure, the RPMs, preparation for takeoff and for landing."

Ruth graduated on August 15, 1944.

"I checked into Wilmington and by the next week, I was on detached duty on Long Island. I flew P-47s, sometimes as many as three a day —delivering them to Newark. Then Gertrude Meserve would ferry all of us back in the C-60 to pick up the next block of planes. So much for Wilmington!"

Maurine "Mimi" Miller (Orr), Class 44-4, overcame several obstacles to become a WASP and, ultimately, a pursuit pilot.[25] Like Ruth Adams, she earned her private license through the Northwestern University CPT program, and then went on to get her commercial and instructor's ratings. When the war came along, Mimi went to Chicago's PalWaukee Airport to instruct Navy flight cadets. "We gave the Navy boys their first thirty-five hours."

Having been told she had a heart murmur, Mimi did not try to join the WASP program. However, she was asked to instruct primary at Sweetwater. After she got there, Dedie Deaton discovered that Mimi, in fact, did not have a heart murmur and suggested that she apply to the WASP. When she entered training, Mimi had 1,213 flying hours, enough to have qualified easily for the Original WAFS.[26]

Mimi was accepted and entered Class 44-6 in January 1944. But given her flying credentials, she was moved ahead to 44-4 and graduated May 23. That made all the difference. She was one of the fifteen from her class sent to a ferrying squadron in May 1944. She was assigned to Romulus, started off ferrying what trainers were available, advanced quickly, and got her time in the C-47—her ticket to pursuit flight. About the time her 44-6 former classmates were graduating at Sweetwater, Mimi entered pursuit school in Brownsville. She graduated on September 15, 1944, and spent the last three months of her WASP service ferrying fighters—mostly P-63s, because that's what the Romulus pursuit pilots were flying.

Class 44-4 was the last class that sent women to the Ferry Command, and only Maurine Miller had sufficient hours and experience to make it all the way to and through pursuit school.

Chapter 28
The Fall Out

With the cancellation of additional WASP training at Sweetwater, Jackie Cochran began to draft a report to General Arnold aimed at achieving the withheld militarization. She had not given up hope. Her report was released to the press August 8, 1944. "Director of Women Pilots Asks Military Status for WASP" is an eleven-page document with a two-page summary.

Laid out for all to see is the wordy but detailed story of the WASP of World War II as told by the woman who coined the name, came up with a concept of how to utilize women in several flying roles, and then sold it to General Arnold.

She does gloss over the WAFS beginnings—a concept quite different from hers. She makes no mention of General Tunner or Nancy Love or the commander of the Air Transport Command, Gen. Harold L. George, whose foresight gave Love and Tunner the latitude to "go" with the initial concept of the women's role in the Ferrying Division. Cochran assumes all the credit. She recounts in great detail what the 1,102 women who flew for the AAF in World War II accomplished, and she makes a solid case for their militarization.

Cochran offers four paths the AAF might pursue:[1]

1. Absorb and militarize the WASP as officers in the lower commission grades;
2. Take no action until an evaluation of the overall pilot situation could be made;
3. Maintain the current civilian program until a decision on militarization could be reached;
4. Because the women were civilians and control over them after training was problematic, the program should be inactivated if militarization was not soon authorized. —To which Cochran added, "If such action should be taken, an effort should be made to obtain military status, if only for one day, and resulting veterans recognition for all who have served commendably."

A prescient plea, given the turn of events.

With the release of the report, what the WASP had been doing for nearly two years was made public. Unfortunately, it was too late.

The press took the "inactivation" option as an "ultimatum" and ran with it. That gave additional fodder to the WASP detractors like the American Legion as well as the uninformed and easily swayed public—those who understood neither the program in concept nor the reality of what the 1,102 women known as WASP did for the war effort. And the saga of the WASP's demise played out in August and September 1944.

Gen. William H. Tunner was reassigned as of August 1, 1944. He left Ferrying Division Headquarters in Cincinnati to take command of the ATC's vital Hump Operation in the China-Burma-India Theater. It was the move—THE assignment—that would cement his reputation as "the man who moved anything, anywhere, anytime," the subtitle of his wartime memoir *Over the Hump*. The airlift he was to command—which over-flew the Himalayan Mountains and the jungles of Burma that lay between India and embattled China—was the lifeline through which the landlocked Chinese Army and the people of China received support and supplies from the Allies during World War II.[2] Tunner's leadership made it work.

His leaving could have been a serious blow to the WASP of the Ferry Command, but it was not. Tunner's replacement was Gen. Robert E. Nowland, who was totally supportive of the women flyers and backed them at every turn. Colonel Robert M. Love—Nancy's husband—was named Nowland's deputy commander.

Three realities emerged following the defeat of the WASP bill.[3]

(1) The Ferrying Division had not pushed for the Costello Bill.
(2) The Ramspeck Committee "took pains to point out that it was in no way attacking the use of the women pilots already trained." Rather, the objection was to continued and increased training of inexperienced young women.
(3) The refusal by Congress to enact the Costello Bill did not conclude the struggle for militarization.

There was still Senate Bill 1810. And after the defeat of the Costello Bill in the House, two groups raised protests. One was the Committee on WASP Legislation, with headquarters in Sweetwater, Texas, home of the Avenger Field flight training school.[4] By summer 1944, the town of Sweetwater was very much committed to the WASP-in-training who had been accepted into their community. The other was the Ninety-Nines—the organization for women pilots. Their efforts were initiated through an article in the July 15, 1944 Ninety-Nines newsletter.[5]

The press remained hostile despite efforts on the part of WASP supporters. Influential columnist Drew Pearson, who jumped the gun on the release date when Cochran's report was to be made public, was particularly opposed to the WASP and wrote a scathing piece that appeared on August 6.[6]

During the long struggle over militarization throughout spring 1944, there had been almost no publicity allowed about the WASP. The Ferrying Division felt that its women pilots, who were doing their job and doing it very well, had not been given a fair deal by the anti-WASP campaign launched by the press and organizations like the American Legion.[7]

Capt. William R. Geddings, Public Relations Officer with the Ferrying Division/ATC, wrote to Nancy Love the day after Cochran's report was released. He had in mind a positive publicity campaign recognizing achievements of WASP in the Ferrying Division. He suggested appealing to women's magazines, because "women's organizations can help the entire program," and encouraging correspondents to make ferrying trips with WASP.[8] Geddings wanted to show that the WASP performed a real service. He suggested photos of WASP picking up airplanes at factories talking with the women who helped build them. He suggested flooding newspapers—from small weeklies to the largest dailies—with hometown news about what their very own WASP was doing.

A *Detroit News* reporter did ride along with Lenore McElroy on a ferrying trip out of Romulus in November and wrote a three-part series about her experience.[9]

The WASP ferry pilots assigned TDY in Farmingdale, Long Island, were about to add one more feather to their collective caps. On September 22, 1944, the corporate and Army brass and the full contingent of WASP ferry pilots gathered to celebrate the production of Republic's ten-thousandth P-47 Thunderbolt.

"They rolled 'Ten Grand' through a paper curtain and out of the hangar door for delivery to the Army. It had already been test flown," Teresa James, the pilot, related. "Jackie Cochran ... was up on a lift-truck platform and she smashed the champagne bottle over the propeller hub. Then they moved the plane back and beckoned me to come on. I was the next pilot up to fly that morning. That's how I happened to be the one who took 'Ten Grand' on its flight to Newark."[10]

As of November 1944, the WASP stationed in Farmingdale had delivered a total of 1,987 P-47s.[11]

On October 2, 1944, General Arnold announced that the Army Air Forces now had sufficient pilots for present combat needs. Arnold said that the total casualties suffered by the AAF since the outbreak of the war were under the numbers originally planned for. This news, following on the liberation of Paris a few weeks earlier, all pointed toward an Allied victory in the offing. The following day, General Arnold announced that the dissolution of the WASP would take place on December 20, 1944.

The need for new pilots had slowed. More veteran combat pilots would be returning to the States now and the reduction of pilot training had made "several thousand male civilian fliers available to be trained and assigned to the routine service pilot jobs now being done by the WASP... Until December 20, current WASP activities will be continued without change."[12]

The WASP received letters from Arnold and Cochran dated October 8, 1944. Detailed instructions for deactivating the WASP were issued on November 1.

That was it.

Since the Ferrying Division had hired original WAFS as civilians—just as it hired male civilian pilots—there was talk that it might be possible to rehire some of the WASP, specifically the much-needed pursuit ferry pilots. The women pilots were well aware of this and talked among themselves of the possibility. When word of this reached General Arnold on October 5, he wired General Giles:[13] "You will notify all concerned that there will be no, repeat no, women pilots in any capacity in the Air Force after December 20, except Jacqueline Cochran."

That same day, the following was sent to the Commanding Generals of all units currently using WASP.[14] "No women will be employed by the AAF in any flying capacity either as pilot, copilot, or member of a flying crew after date of WASP inactivation on 20 December 1944." — signed Arnold.

That edict stood for more than thirty years.

On November 1, General Nowland, Ferrying Division commander, made a last-ditch effort on behalf of his women ferry pilots. The letter he wrote to his boss, General George, had the word SECRET emblazoned across the top.[15]

The cost of retraining male pilots to replace the WASP pursuit pilots upon their deactivation would be $9,336 per man, Nowland wrote. The total cost of training 117 men to replace the 117 women pursuit pilots leaving as of December 20, 1944, was calculated at $1,085,312.

"Replacement of these experienced [women] pilots with an equal number of male pilots will entail a month's training at Brownsville for each male pilot. These replacements must be chosen from the ranks of our more experienced personnel since it has been demonstrated that a minimum of 400 hours and an instrument card are necessary for pursuit training and subsequent safe delivery of fighters," he said.

The plan had been to keep the civilian women pilots on pursuit deliveries permanently, allowing the release of male pilots for progressive upgrading "in order to meet our requirements." And he added that those 117 officers would, of course, be ineligible for upgrading for a minimum of five months. After the month of training, each male pilot would need to spend at least four months of pursuit ferrying to achieve the minimum period to ensure effective utilization of his training.

It was a waste of taxpayers' money, but worse, it meant sidetracking 117 men for a period of at least five months to do a job that 117 women were already trained to do and were performing quite admirably. But in the political climate of 1944, it did not matter.

Thus for the WASP began the unraveling of a dream.

Chapter 29

Ferrying — the Highs and the Lows

Ferrying put on a new face in September 1944 as the remaining WASP ferry pilots continued to crisscross the country via the airways.

New to the WASP were two high-performance, twin-engine pursuit aircraft—the P-38 and the P-61—though deliveries of these affected only an elite few among the women pursuit pilots. Opposite of these shiny new aircraft were the shot-up war wearies that had been brought back from combat and the beat-up, student-abused trainers, all of which had seen their war and were now destined for extinction. Dubbed "Class 26," these aircraft were barely flyable.

The P-38—with two 1,475-horsepower engines and a maximum speed of 415 mph at 25,000 feet—was built by Lockheed in the LA Basin.[1] Long Beach ferry pilot Iris Cummings's logbook shows a continuous string of delivery flights in the P-38 throughout the month of October and into November.

Photo 38. Iris Cummings, Class 43-2, Long Beach ferry pilot who flew the P-38.

Photo courtesy of the Iris Cummings Critchell Personal Special Collection

How many WASP actually ferried the P-38? The WASP exhibit at the P-38 Museum at March Field in California—a joint effort put together by WASP P-38 pilots Iris Cummings Critchell and Mary Lou Colbert Neale —lists twenty-three. Critchell and Neale sought actual orders as proof when putting their collection together. In her book, *Women in Pursuit,* WASP Kay Gott—herself a P-38 pilot—lists thirty-one women who were Class IV pursuit pilots and qualified to fly the Lockheed Lightning P-38.[2]

Dallas WASP squadron leader, Delphine Bohn, wrote the following about her unforgettable flight to Fresno, California, in a Lockheed F-5E photographic version of the P-38:[3]

> Crossing the Tehachapi Mountains, the last checkpoint before letdown, I saw Fresno's field and runways despite its highly touted camouflage. On the radio I identified my aircraft and myself by the proper wartime code for entry into that vulnerable and highly "skittish" West Coast area. When I lifted my head out of the cockpit, the field had vanished! No airport, no runway.
>
> Already I was down to pattern altitude. I should long since have seen the field and contacted the tower. Finally the tower called. "Your position, please?"
>
> In a hesitant voice I answered, "I don't know."
>
> I was informed that I was on final, to prepare for touchdown. Then, only because my wheels rolled on hard surface, could I be certain I was no longer airborne. At end of rollout, not one but three jeeps awaited me.
>
> I eased in after the "Follow Me." The two remaining jeeps, exceedingly well manned, eased in behind my P-38, single file. At the line, I shut down. My escorts stopped surprisingly close, one on each side. Several enormous, very well armed males—military police, for that was their obvious capacity—stepped down from each. They formed a line to await my descent from the P-38.

> When I managed to climb down, it was made crystal clear to me that I had used the wrong wartime identification code. I'd used the code that alerted the entire West Coast to an enemy invasion.

Photo 39. Dottie Kocher, Class 43-4, Long Beach ferry pilot, standing on the wing of a P-38.

Photo courtesy of the Iris Cummings Critchell Personal Special Collection

Dorothy Kocher Olsen recalls, "We could fly the twin-engine pursuits at night. I'll never forget my night checkout, flying a P-38 under a Texas moon so bright that I could read my map without lights. I never saw a night like that. In the hours after midnight, I listened to beautiful music from the command radio center—all the way across Texas.

"I was going to take it just to Palm Springs—to see what it felt like—but I kept going, all the way to Love Field. I got there at three o'clock in the morning."[4]

The other distinctive twin-engine, twin-boom aircraft—the P-61 Black Widow—was built by Northrop, also located in the LA Basin. It was the first operational U.S. military aircraft designed specifically for night interception of aircraft, and was the first aircraft specifically designed to use radar.[5]

"It was a remarkable airplane—a bigger mass of airplane than a P-38, with two 2,000 horsepower radial engines," Iris Critchell recalls. "It was massive and painted black, with this big hump right in the middle of the airplane—where it carried the first airborne radar—and a red light in the back. It looks like a Black Widow spider—that's what it was called. It was a single-pilot cockpit with the radar operator's compartment in the back.

"Radar was still so new—it was secret. Everything about the airplane was rated confidential. As a result, when we flew it, we were not supposed to tell anybody what we were flying or where. And we were required to carry the .45 automatic with us, because this airplane had to be guarded. If we RONed while flying one, we had to have a guard, with rifle in hand, marching up and down right in front of it, all night!

"I loved that airplane, because it really was designed by a pilot for a pilot. Instead of putting all the levers where you can't reach them, you had everything where you could reach it. And it was logical to have the control handles where they were. It was a wonderful privilege to get to fly that airplane."[6]

Few WASP got to fly this one. Betty Gillies flew it first, and the rest were all from Long Beach near the factory's location: Barbara Erickson and Iris Cummings as well as Katy Loft and Ginny Hill (both 43-4).

"When Betty Gillies came to Long Beach to check out on the P-61, she stayed in the barracks overnight," Iris recalls. "We weren't supposed to tell anyone what we were flying or where. But she told me, 'I've sent a telegram to Bud [her husband] and told him that tomorrow I'm spinning a web to Sacramento.'

"Bud would know instantly what airplane she was flying, when and where she was going. Betty always sent Bud a cryptic telegram like that, letting him know that—wherever she was—she was in and safe for the night."

Iris was checked out in a P-61B at Long Beach, on November 9.

Ginny Hill earned her twin-engine pursuit rating in the P-61, unusual because the P-38 normally came first on the transition list. "I checked out in a P-61 because I was tired of sitting around the base grounded until I got my twin-engine pursuit rating. All the P-38 trainers were mechanically out of service. After hanging around for two weeks waiting for a P-38 to get my time in so I could get back to ferrying, I checked out in the Black Widow and delivered several of them to the East Coast."[7]

And then there were the Class 26 aircraft, nicknamed War Wearies.

Kay Gott writes, "War-weary airplanes sometimes fell apart as they were being flown to depots. ... At the end of 1944 to ferry a war-weary airplane was very 'chancy,' not highly regarded by any ferry pilot as a desirable job."[8]

Cappy Vail was given a very old SBC-6 (A-25), a Navy Curtiss model, to deliver. "They wanted me to fly that from Dallas to Long Beach. I stopped at El Paso Airport, refueled and took off again. Well, nothing from the hydraulic system would work. The landing gear didn't come up, so obviously I checked the flaps and the other things that we use the hydraulics for and none of them worked. They had me fly around burning up the gas before I could land.

"I looked down at this short runway and right over on the north side was this great, big beautiful Army Air Forces Base with a long runway. I asked to land there because, without brakes, I would just coast to a stop. They said no, I had to land at El Paso because the contract was with them.

"I made a very fast landing at El Paso without flaps or brakes. I saw the tower whisk by me. I didn't know where the end of the field was. I had no more reference points. So I just unlocked the tail wheel and the plane did a ground loop. But it didn't drag a wing because it was a former Navy plane and had this wide landing gear—you know, for landing on a carrier. I didn't even blow a tire."[9]

"Flying war-weary junk airplanes gave me more than a few gray hairs," Teresa James related. "I picked up an A-25 out West and nursed that baby all the way to the east coast. First, I couldn't get the gear up. Then the radios went out. Next, I couldn't get the bomb-bay doors closed.

"Approaching Aberdeen, Maryland—my final destination—exhaust fumes poured into the cockpit and I had to request a straight-in approach. I had to land with the canopy open and the wind blowing in my face. It was hard to see the exact height above the ground.

"I walked into Operations to get my Memorandum Receipt signed for the aircraft and told them how much trouble I had getting there in one piece.

"The Operations officer said, 'Well, no wonder. It's a Class 26 airplane.'"

"What does that mean?" Teresa asked.

"They'll taxi it out to the Proving Ground and use it for bombing practice. The pilots will bomb the hell out of it to find out how accurate their air strikes are."[10]

Betty Gillies flew a lot of war-weary P-38s the latter part of 1944. "These were airplanes that had been sent back from the European war zones, and they were very, very tired," said Betty.

Flying one out of Harrisburg, Pennsylvania, she had an engine quit on her on takeoff—the very airport where Evelyn Sharp had died a few months earlier when the left engine of the P-38 she was ferrying quit. The wind was from the opposite direction the day Betty took off. She flew out over the river rather than having to climb to get over hills and she was able to maneuver the airplane safely back on the ground.[11]

"One time I was asked if I would mind taking this BT-13 to Death Row," Marie Muccie recalled. "What's that?" I asked.

"They said, 'Well, they're going to take it apart and junk it.'

"Why don't you just junk it now?" I asked. "Take the wings off and send it there in parts."

"'No,' they said, 'you have to fly it there.'

"That plane was not flyable. I risked my life flying that piece of junk. Every time I'd get in it and take off, it would lose manifold pressure. So, I'd fly to the next airport and have another mechanic fix it. It should have been an easy flight. It took me five days! They'd repaired it just enough for me to get to the next stop.

"Finally I delivered it. That was actually the only plane that I ever flew that wasn't brand spanking new, and I loved the brand new ones."[12]

In November 1944, Jill McCormick (43-5) was one of several Wilmington pilots on a delivery of war-weary A-24 dive bombers out of Baltimore headed for the reclamation center at Reading, Pennsylvania. The engine on her Douglas Dauntless exploded and burst into flames after she made her climbing turn over the harbor and headed toward Reading. Jill had learned ditching procedures in pursuit school. She put the aircraft into a dive but pulled out just before she hit the water, skimming the surface. The spray put out the flames and the plane came to a stop.[13]

Photo 40. WAFS Nancy Batson had a scare when the nose wheel on the P-38 she was ferrying would not go down and lock.

Courtesy WASP Archive, Texas Woman's University, Denton, TX.
The Wilmington ferry pilot persevered and got both the nose wheel and the P-38 down very much intact.

Flight leader Nancy Batson described the incident from her perspective. Circling above, counting all her chicks as they took off and rendezvoused with her aloft, Nancy saw, “out of the corner of my eye here comes a boat tearing along the shore. I looked down and there sits Jill McCormick, perched on the wing of her airplane and the airplane is sitting in the water off the runway.”[14]

And then there were the lemons.

Nancy Batson finally landed her troublesome new P-38 in Newark after twice having to blow the nose gear down with the emergency CO^2 cartridge—once in Pittsburgh and the second time in Newark, where she greeted the line crew with, “boys, this time I’ve brought you a real lemon.” [15]

Jane Straughan vividly recalled a P-63 she took from Romulus to Oakland. “The electrical system kept going out. I stopped in Coolidge, Arizona. They worked on that plane for two or three days. When I was ready to take off, the engine caught fire. More repairs. More delay.

“From there, I got to Palm Springs. The thing was in the shop for four more days. Finally, I landed at the base in Oakland. The Operations Officer took one look at the papers and said, “what in the world!”

“It’s a lemon,” I said.

“He told me I needed to take it to the shop across the bay.

“‘I’m not flying that across the bay,’ I said.

“ ‘I’m not letting one of my men fly it over there,’ he said. ‘It might go down.’ ”

“My point exactly!”[16]

Ruth Adams was in South Carolina with a group of male pilots waiting to take P-51s to Newark. “I got about fifty feet off the ground and the plane lost power. My heart was in my mouth. I had given it full throttle because that’s what the instructions said, but I eased back and the engine caught.

"Signing in at Newark, I told the pilot standing next to me about my terrible experience taking off from South Carolina. He said, 'you know, the same thing happened to me. It must have been water in the gas or something. Was never so scared in my life.'

"He was an ex-combat pilot!

"I was never taken in by the glory of flying pursuits. I was never unaware of the danger."[17]

The engines in the P-51s were liquid cooled. The coolant temperature gauge was not supposed to exceed 124 degrees F. If there was a danger of that, the pilot was to shut off the engine, because the coolant valve could blow open, or in pilot jargon "pop off." And that meant a tow truck had to bring the aircraft in.

Viv Cadman had one blow up in El Paso. "I had landed, and it was hot. I had to taxi around and to get back to the parking strip, I had to cross a runway. It was a crazy situation. I was sitting there getting hotter and I said, 'Permission to cross.'

"I was told to hold. They had somebody landing. The coolant foamed and I slammed the canopy shut just before it streamed all over. I'd have been scalded. I shut everything off. They had to bring out the tow truck."[18]

Part VIII

Denouement

November–December 1944

The U.S. Navy defeated the Imperial Japanese Navy in the historic battle of Leyte Gulf in the Philippines, October 26. It has been called the greatest-ever sea battle.[1]

Franklin Roosevelt won an unprecedented fourth term in the White House November 8.

General George Patton's troops crossed the Moselle River on November 9 and the Saar River December 5 as the march toward Berlin moved two steps forward.[2]

December 16, American band leader Maj. Glenn Miller's airplane was reported missing over the English Channel. His popular band was scheduled to play a concert in Paris on December 21.[3]

The WASP of the Ferry Command were devastated, but they flew on. They also prepared for deactivation. The date of their departure was December 20. They would be home in time for Christmas.

Chapter 30

Twelve WASP Ferry Pilots Died

The WASP and their leaders were fortunate that no lives were lost ferrying Class 26 aircraft. But as of September 1944, Nancy Love and the Ferrying Division were still to feel the sting of death among their own three more times before deactivation.

Alice Lovejoy (43-5), stationed at Romulus, was sent to pursuit school in Brownsville early in September 1944. She died September 13 when another plane collided with the AT-6 she was flying during training.

Hazel Ah Ying Lee (43-4) graduated from pursuit school October 1, 1944, and returned to Romulus where she joined in the constant movement of P-63s to Great Falls, Montana. On November 23, approaching Great Falls, her P-63 and another collided and burst into flames. Dragged alive from the wreckage of her plane, Lee died two days later, November 25, 1944, as a result of her burns.

Her 43-4 classmate, Dallas ferry pilot Virginia Luttrell, was an eyewitness.[1]

Photo 41. Virginia Luttrell, Class 43-4, Dallas ferry pilot.

Courtesy WASP Archive, Texas Woman's University, Denton, TX.

"I was with Irene Fremd (43-5) and two men from Dallas—Jeff Russell and Charlie Dyer. The four of us had been together on trips before. We had delivered 51s to Newark and were sent to Buffalo to pick up P-63s. We sat around Buffalo waiting for the weather to clear. Finally, we flew on to Madison, where we sat again and waited.

"In the meantime, people were piling up behind us. We went to Fargo and sat there for another two or three days. By this time there must have been at least forty to fifty airplanes, all P-63s, piled up in Fargo. At last, on Thanksgiving Day, the weather cleared enough for us to take off. So Irene, Charlie, Jeff, and I took off sort of as a group.

"Shortly after takeoff, Jeff moved over close to me and pointed to his earphone and raised his hand. His radio was out. I motioned to him to get behind me and stay there. And he did.

"This is the way we flew into Bismarck. When we landed there, I called the tower and said that there was a P-63 behind me without any radio and would he please give him a green light. I gave him my position and said, 'I am on final approach; the plane right behind me has no radio. Turn on the green light so that I know you understand that he's behind me.' He gave me the green light and I guess Jeff waggled his wings to accept it. We came in and landed safely.

"After all four of us got back together again, Jeff went over to find out if there was some way to get his radio fixed. Because it was Thanksgiving Day, there was nobody. Somebody was there to gas the planes and that's all. Nobody to fix the radio. Irene told Jeff to follow her into Great Falls. Charlie and I took off, then Irene, and then Jeff behind her.

"When we reached Great Falls, the air was filled with P-63s circling. I went around a couple of times before I was finally cleared to land. Then Charlie landed. Pretty soon Irene came in.

"We said 'where's Jeff?'

"She said he should be on final approach.

Photo 42. Romulus pilots A.J. Starr, 43-4, and Fran Snyder, 43-6, sitting on the wing of a P-63 with Hazel Ah Ying Lee, 43-4, standing below.

Courtesy WASP Archive, Texas Woman's University, Denton, TX.
The photo was taken in Niagara, NY, in late November 1944 when all three were ready to take off in P-63s on what became the fateful flight to Great Falls where Hazel was killed.

"We looked and here comes a plane making the proper approach, the short approach. Way out there behind, is a plane dragging in almost flat. The two planes got right together at the end of the runway. The plane underneath pulled up into the one on top and they both crashed right there at the end of the runway. Hazel was in the bottom plane and Jeff was in the top plane.

"We found out later that the tower had lost track of which plane had no radio."

"Jeff was coming around on his final approach and Hazel was dragging in. When the tower saw what was happening, they said 'pull up, pull up.' But the only plane that could hear it was Hazel. And she pulled up right into Jeff—who, of course, heard nothing.

"The plane caught fire. Rescuers pulled Jeff out. His foot was burned. Other than that he wasn't hurt.

"Hazel was trapped in the bottom plane. They dragged her out. Very badly burned, she died two days later. She was conscious the entire time. The doctor said they had never seen anyone so brave. She never complained. It was a very, very sad thing."

Hazel was one of two Chinese-American women to fly for the WASP. The other was Maggie Gee of Class 44-9.

The disappearance of WASP Gertrude Tompkins Silver (43-7) is a very different story. She, too, graduated from pursuit school October 1, 1944, and then returned to the 5th Ferrying Group in Dallas. On October 26, she departed Los Angeles Mines Field in a P-51, part of a flight of P-51s bound for Newark. She never arrived at her next scheduled stop, Palm Springs. Sea, air, and ground searches revealed nothing, but they were late getting started because the Ferrying Division did not know for a few days that Gertrude was missing.

The overcast that usually hung around Mines Field (now LAX) until well into the afternoon meant that pursuit pilots frequently had to take off and then fly through a few hundred feet of scud clouds before breaking out into the sunshine everyone knew was just above the layer of crud.

Vivian Cadman flew many P-51s out of Long Beach and she had a theory as to what happened to Gertrude.[2]

"When we were ferrying the P-51s, we'd be over at Mines Field all morning long waiting to take off. By the time we'd get off the ground, we were tired and by then we were flying into the sun. I remember talking to myself, 'don't look out.'

"We took off out over the water and we'd be on top of the murk within a minute or two. But before we climbed out of it, it was just awful to look out and see nothing. I would stay glued to the instruments, watch that altimeter, keep that artificial horizon straight, and climb, climb, climb. Then I'd break through and be on top and it's blue. I'd take a breath, make a 180 and head for Palm Springs.

"I think Gertrude looked down and got vertigo. The end of the runway is straight out over the water. I think she looked out and she went straight in—right off the end of the runway.

"They didn't even miss her for three days. *Three days*! Nobody ever reported her missing."

Virginia Luttrell, who was stationed with Gertrude in Dallas, again was an eyewitness—though of a different sort—to Gertrude's disappearance.[3]

Virginia had just reported to the Ready Room at Love Field to pick up a P-51. "The squadron leaders Ann Johnson and Delphine were there, and Gertrude Tompkins's husband. His name was Silver. They had only been married a couple of weeks. He came down to meet her because the flight was due in that day. When I got there, they were just checking to find out when she had last RONed. Evelyn Tomlinson (43-6) also was on that flight. The clerk in the RON office said, 'I've heard from Tomlinson.'

"They said, 'what about Tompkins?' There was no RON from Tompkins. All they knew was the flight of P-51s had taken off from Mines Field very early the morning of October 26. Destination Newark.

"Now I've taken off in a P-51 from Mines Field. They pay no attention to the weather. Restrictions? If we had to fly a P-51, didn't matter. If you could see the airplane, if you could see ten feet in front of you on the runway, you took off. The other girls on that flight said that they couldn't see the end of the runway, but they took off.

"Their first stop was Palm Springs. Gertrude wasn't there, but there were so many of them that either they were not aware that she wasn't there or they thought she was lagging behind. They were on this flight two or three days. There was absolutely no record of Gertrude. It's as if she vanished into thin air.

"Whether she became disoriented after takeoff, who knows," Virginia said. "But I cannot believe she became disoriented. In order to fly a P-51 you had to have an instrument card and you had an altimeter. The engine may have quit on takeoff and she went into the ocean. The only other possibility is that she made the 180-degree turn, headed for San Gorgonio pass, and ended up in some of those rugged mountains up there. There are those two possibilities. The guy in the RON office had a TOM on his list for Tomlinson, and that one TOM is all he paid any attention to.

"I can still see the stricken look on her husband's face. He could not believe this was happening. Neither could anyone else. Delphine called everybody involved, and she was really upset. Gertrude had disappeared."

Searches for Gertrude's P-51 continue today, but her fate still eludes her family and the others searching for her.

Gertrude and Hazel were the last WASP ferry pilots to perish, but there were other WASP deaths in late fall 1944 right before the program ended. A total of thirty-eight WASP gave their lives flying for their country in WWII. Twelve were pilots with the Ferrying Division when they died. Four died ferrying pursuit aircraft: Evelyn Sharp (WAFS),

Dorothy Nichols (43-2), Gertrude Tompkins Silver (43-7), and Hazel Ah Ying Lee (43-4). Two, Dorothy Scott (WAFS) and Alice Lovejoy (43-5), died in training at pursuit school. One, Marie Sharon (43-4), a Long Beach ferry pilot, died in the crash of a B-25 while attending instrument school in St. Joseph, Missouri. Both she and her instructor were killed during a night instrument training flight in severe weather conditions. Susan Clarke (44-2), Cornelia Fort (WAFS), Paula Loop (43-2), and Virginia Moffatt (43-2) all died ferrying BT-13/15s. Mary Elizabeth Trebing (43-4), died ferrying a PT-19.[4]

Texas Woman's University, the home of the official WASP Archives, published *In Memoriam: Thirty-Eight American Women Pilots* in 1993. This booklet tells about each of the thirty-eight. Four WASP worked on the booklet, preparing it for publication: Dawn Seymour (43-5) memorials chair, Clarice Bergemann (44-2), Jeanette Jenkins (44-1), and Mary Ellen Keil, (44-2). The printing was made possible by a bequest from the estate of Joann Garrett (43-6), and gifts from WASP families and friends.

TWU has plans to update the booklet.

Chapter 31

Facts of Life in the Ferry Command

In her final report, Jacqueline Cochran wrote that, as in the case of male pilots, some women pilots are better than others—suitable for one type of work and not for another. The yardstick by which the WASP training program should be measured was NOT if the graduates could fly pursuit aircraft or B-17s. The measure of success was: (1) could they safely ferry trainer-type airplanes under daylight, non-instrument conditions; and (2) could a significant proportion of them qualify for transition into higher classifications?[1]

The original training program was built on that premise. However, by fall 1943, those goals no longer meshed with the Ferrying Division's needs because the FD's mission had been altered. Now it needed to produce male pilots who could fly four-engine bombers and transports overseas. And beginning fall of 1943 on, there was an increasing logjam of male Class I and Class II pilots. For the "good of the service," only the most capable women pilots could be transitioned to fighter ferrying.[2]

Yes, the dynamic changed mid-stream—through no fault of Nancy Love or Jackie Cochran. The war and the necessity of winning that war dictated the circumstances and the realities.

Nancy Love had told Jacqueline Cochran—in their July 3, 1944, phone conversation—that unless the women sent to them could be quickly qualified for pursuit school, the Ferrying Division could not use them. Cochran understood that and placed subsequent graduates accordingly.

The Ferrying Division was desperate to hold on to its women pursuit pilots and to keep them in the air. The ferrying of pursuit aircraft was crucial to the war in Europe. P-51s were needed to protect bombers flying into Germany. Men were needed to fly multi-engine bombers and cargo aircraft abroad, so every woman who ferried pursuit aircraft to the docks in the U.S. freed a man to fly aircraft overseas and also put a pursuit aircraft on a ship bound for England.

But with the defeat of the WASP militarization bill, the elimination of any new training classes at Sweetwater, and the knowledge that sentiment against the WASP's usefulness had changed, the program was winding down.

The women of the Ferrying Division were flying their hearts out, but the strain was taking its toll. Several were ill or rundown. Nancy Love spent the first week of September 1944 at Love Field filling in for her ailing squadron commander Delphine Bohn, who was on medical leave recovering from pneumonia. On September 4, Love wrote to Delphine and summed up her state of mind.[3]

> The WASP situation remains in its usual SNAFU state, but I've acquired a new slant on it, and have tried to transmit same to the gals, i.e., we are now the long-sought-for, small select group. We've got a heck of a good record behind us and an even rosier one ahead, or should have with this bunch of good pilots. So forget the newspapers, the Training Command, J.C., WTS/CAA and the

> whole damned mess—and fly! ... Fly the pants off any and all airplanes assigned to us—from L2s, to A-20s or B-17s. ...
>
> When you get back, you'll find your squadron small and self-sufficient, and there's no reason squadron leaders can't get out and have fun—only one responsible person need be there. We've all gotten so violent about this that I think it's broken down our health—you and Esther Manning with pneumonia, I've had flu and now have trench mouth, of all things. So let's stop worrying and have fun—and see what happens.

The pervasive stress took its toll on Wilmington's Helen McGilvery (WAFS).[4] In June 1944, Helen attended Officer Training School in Orlando. She graduated June 30. Helen's flight records show that she resumed flying in Farmingdale July 6, with subsequent ferrying flights daily through two P-47 deliveries on July 14. She has no flights recorded the rest of July or in August.

Something happened.

In Helen's papers is a personal, handwritten letter from Betty Gillies mailed July 17, 1944. Helen had told Betty she wished to resign from the WASP. Betty writes:[5]

> Got your letter yesterday. ... Gosh but I was sorry to read the message it contained. Do you really have to do it, Mac? I tried and tried to read between the lines, to try and see why—but got nowhere. Sure wish I had a chance to talk to you about it because if there is anything I could do to prevent it, I sure would do it! I don't, of course, mean that I would prevent your leaving if you really think it best, but if there is any maneuvering I could do that would make things work out better for you so that you could stay in, I sure would try.
>
> ... How about holding up your decision to resign until the future plans for the outfit unfold? However, if your mind is really made up, it will be necessary for you to submit a regular formal letter of resignation through channels.

> ... Sure hope you'll change your mind and stick with us. There is going to be a heck of a lot of work to do and we need you, Mac.

A memo in Helen's file dated September 5, 1944, states the following: "Attached is a Memorandum, Subject: Flying Safety and Court Martial Sentences for Violations of Flying Regulations." Helen signed it and a second September 5 memo that states: "I certify that I have read and that I understand Memo 55-65 dated 10 July 1944, and placed same in my personal file."

Helen had made several P-47 deliveries between July 10 and 14, but obviously had run afoul of a flying safety regulation and was called on the carpet for it. Nancy Batson and Teresa James, two of her closest friends, related in interviews that Helen was upset that summer because she believed that her husband had been unfaithful. Helen was devastated. For whatever reason, she wished to resign.

Helen was one of four original WAFS who came down with pneumonia over that long summer of 1944. The other three were Dorothy Fulton, Delphine Bohn, and Esther Rathsfelder. Helen spent all of August with her husband's family on Long Island recuperating under the watchful eye of her physician father-in-law.

But she responded to Betty's plea and did come back because her flight records show that she commenced flying again in September, delivered her first P-38 on September 17, 1944, and continued to serve until the WASP deactivation December 20, 1944. Her final flight as a WASP was to ferry a P-47 from Farmingdale to Newark on December 17.

There were other resignations to deal with.

In her final report, Cochran points out that seventy-three women had resigned from the WASP by August 1, 1944, when the report was dated. The author's rough count indicates that fifty-one women resigned from the Ferrying Division over its full twenty-eight-month existence.[6]

Reasons for leaving included marriage, pregnancy, grounded for medical reasons with little chance of regaining flight status, illness in the family, the need to care for a child or some other family member, objections by a husband to his wife's flying, desire to join another service, or simply the desire to leave for personal reasons. The greatest number listed getting married or being pregnant as the reason for leaving.

"Ferrying was a very tiring service because they had so many planes to move," Marie Muccie (43-2) recalled. "We had no time to ourselves. When they trained us women to do those jobs, we released men for combat. [The men] did not want to go to combat. That's why some of the men resented us, and my husband was one of them. He didn't like the idea of the WASP. 'You shouldn't be flying airplanes. A woman's place is in the home,' he would say. We had fights all the time about that."

When she was still single, Marie was flying out of Fairfax—copilot in the twin-engine B-25s, which fly in any kind of weather. She needed her instrument rating, so Betty Gillies sent her to instrument school. Captain Mike Genaro was the head of the instrument school. He became Marie's instructor.

She did well, but he refused to pass her. So Marie confronted him and threatened to report him. He passed her. She got her instrument rating, but in the meantime they had become romantically involved. "After I got my rating, I wished I didn't have it because I never liked instrument flying anyway. You're up there and you can't see anything. I like to see out of the airplane and see where I'm going. One mistake and you don't know where you are.

"I married Captain Genaro in the service. He said, 'I don't want you to stay in the service and fly anymore of these planes.' At that time I had had so much flying, I was totally exhausted. I didn't want to look at another airplane. I didn't want to fly anymore. I was ready to give it up. We were able to resign, so I did leave in August 1944. We knew we would be disbanded anyway."[7]

Marie wasn't the only ferry pilot to leave early because of a husband's request. Enid Fisher (43-6) was ferrying out of transition-focused Long Beach, and climbed the ladder fast. She reported mid-October 1943, managed to qualify for pursuit school by the end of May, and graduated on June 30, 1944. She joined the constant stream of Long Beach P-51 pilots making their way cross-country from LA to Newark. But her husband of three years was not happy with her flying those pursuits. Finally, she gave in. "I resigned from the WASP in October 1944 at the insistence of my husband. As Congress said, we had a surplus of military pilots and we were to be disbanded in December 1944."[8]

Some had regrets.

One WASP pursuit pilot, in an interview years later, stated that she had made a mistake leaving to get married before the WASP were disbanded. She still agonized over her actions and told the interviewer, "I feel like a quitter."

The sad part was, in conversation with a friend not long before the wedding, she realized her mistake. Her friend—catching on—convinced her to call her WASP squadron leader to see if she could still come back, but the squadron leader was out on a trip.

One of her WASP buddies managed to fly in for the wedding. When the bride saw her friend, dressed in her WASP uniform, "I nearly turned around and walked out of the church."

She went through with the wedding, but carried her regrets the rest of her life.[9]

Margaret Reeves Riviere (43-4) was a stalwart in Class I, II, and III deliveries. Her logbooks indicate that she was constantly out on trips, ferrying increasingly larger aircraft out of Wilmington. She logged deliveries in everything from a Cub – the first she had ever flown—to UC-78s and an A-24. She did not, however, opt to go to pursuit school, though she was sent to instrument school, July 7, 1944.

Margaret was not included in the Great Transfer, but she left the Ferry Command the end of August to go home and take care of her aging father. She received the following letter from Esther Rathsfelder, Wilmington Executive for WASP Operations, dated September 3, 1944.[10]

Dear Margaret,

> Enclosed herewith is your paycheck for the month of August 1944. In all the commotion of your departure, I don't guess I got around to telling you what a swell job we all think you have done with the WASP of the 2nd Ferrying Group. I don't believe there is any girl in our squadron who has come in from a trip and gone out on another one on the same day, sometimes within an hour, more often than you have. Despite that, you never complained, just gathered up your goods and chattels and the girls in your flight and went of again. In the more than a year that you have ferried, you have a perfect record of trips, as quick as was consistent with safety, and no accidents. Your capabilities and cooperative attitude have been appreciated by everyone concerned and will certainly be missed in our group. We are all sorry that you could not stay with us any longer. Please let us hear how you are getting along once in a while. Hope all goes well with you and that you had a good trip home.
>
> Sincerely yours, Esther M. Rathsfelder

Esther Rathsfelder, herself, left the first of October following her bout with pneumonia. Esther had served Betty Gillies ably for two years.[11]

Inez Woodward Woods (43-4) did accept a transfer to the Training Command. She loved ferrying, but did not want to fly fighters. "My background at Curtiss Wright, before the war, entered into that. I knew that flying fighters was probably the touchiest of all flying. Your chances were the poorest. So I didn't volunteer to get into fighter training."

In August 1944, Inez was transferred to San Marcos, Texas, to fly navigation cadets, primarily in twin-engine AT-7s. She remained there into October when she went home to take care of her ailing mother.[12]

Barbara Ward (Lazarsky), 43-4 from Romulus, upon reading Drew Pearson's opinion of the WASP, "concluded that my services were not essential."[13] She got married and resigned.

Helen Stone (43-2), Wilmington ferry pilot who had heard that Cochran considered WASP who resigned as "ungrateful," wrote that she had come to the conclusion "that there are now far more pilots than are needed and so long as the shortage is over it is time to resign."[14]

Four who married in service and stayed in were pursuit pilots Cappy Vail—whose story is told in an earlier chapter—Marty Wagenseil, Gayle Bevis, and Virgie Jowell.

Virginia "Virgie" Jowell (43-4) met her husband-to-be, Jim Hagerstrom, when she was in Orlando for OTS, spring 1944. Jim was stationed in Orlando and they met at a dance. "He told me later, he went back to the barracks that night and told the sergeant there that he had just met the gal he was going to marry." Immediately after OTS, Virgie was sent to pursuit school in Brownsville. She graduated July 15. In the meantime, Jim had contracted malaria. While he was on sick leave, they decided to get married.

At that point, Nancy Love was setting up the Evansville sub-group to ferry the modified P-47s. "Jim asked to be transferred to the P-47 Republic mod center in Evansville. When I told Barbara Donahue that Jim was being sent there, she put me on the list for Evansville. Jim and I got an apartment in Evansville and I ferried out of there until deactivation. The only difference was, I was transferred to the 2nd Ferrying Group so, when the photo of the final party—December 19 in Romulus—was taken, I wasn't in it. I was in the Wilmington photo instead."[15]

Gayle Bevis graduated from Sweetwater in September and married in December 1943. "People used to ask, 'Why aren't you with your husband?' Well, he was a basic flight instructor in Enid, Oklahoma. There was never any question. We saw each other as much as we could —weekends or trips. One time, I faked a mechanical on an airplane so I could land in Enid overnight. Little things like that. Or we would meet at Little Rock, Arkansas."[16]

Martha Wagenseil and Capt. William H. Davis Jr. were married August 12, 1944, in the Base Chapel at Long Beach. They had met in April at pursuit school in Brownsville. Now Bill was going overseas. "We felt that we wanted to belong to each other if we were to be separated." Martha continued to ferry P-51s out of Palm Springs. Bill "came back alive from Assam in India and China." Like most of the male ferry pilots, he had put in his time flying "the Hump."[17]

Jane Straughan and Al, her husband of several years, both were ferry pilots. She describes an unplanned meeting in North Carolina. "I ran into my ferry-pilot-husband, Al. Well, of course, we were delighted. There were bunches of ferry pilots coming in and going to the hotel. Al took the parachutes and our B-4 bags and I went to register. This very nice clerk behind the counter, an older man, asked me if I would step aside. He advised me that they didn't like to register male and female pilots in the same room.

"I said, 'Well, this is my husband.'

"'But we prefer that you do not register.'

"I said, 'We are *married*!' I was getting huffed.

"Finally I got out my orders and Al's orders to prove our last names were the same. He was very apologetic, because in the meantime he had explained to me about venereal diseases. I still was huffed, but then I thought, 'This *is* funny.' When Al came over, of course, he thought it was funny. He didn't see why I was huffed at all.

"Well, there were two older women getting on the elevator. I guess they had seen Al come over to me, and we got on the same elevator. One of them said, without lowering her voice, 'Thank god that's not my daughter over there. You don't know what they'll be doing now.'

"I said, 'Al, gee, this is really nice because now we've got to live up to our expectations tonight.' "[18]

Several WASP were war widows. Some entered service as a result of widowhood and others, widowed while in service, kept flying to remain active and involved rather than resign, go home, and feel sorry for themselves.

Dorothy Webb, 43-6, was widowed while she was in training in Sweetwater. When she graduated, Dedie Deaton sent her to Long Beach—close to Bakersfield where her parents were caring for Dorothy's daughter. Between October of '43 and May of '44, Dorothy worked through the transitions and was rewarded with a slot at pursuit school in Brownsville. She graduated June 30, 1944, and returned to Long Beach to ferry pursuit aircraft until WASP deactivation.[19]

Original WAFS Teresa James, the third woman to fly the P-47, was a fixture in Farmingdale flying the daily runs from the Republic factory to Newark daily. She was on duty in late June 1944 when she received the dreaded telegram—*Missing in Action.*

On June 22, sixteen days after D-Day, her husband George "Dink" Martin's B-17 was hit over northern France and had not returned to base. He was classified as missing on an operational mission. Teresa went home to Pittsburgh. Two days later she received a call from Betty Gillies. "We need you, Teresa."

"I'm better off flying and being with the girls," Teresa told her. "I'll be back tomorrow."

Her husband never came home.[20]

Detailed instructions for deactivating the WASP were issued on November 1, 1944. Each WASP in good standing would receive a certificate of service and certification of pilot qualifications and horsepower rating. The CAA would consider certified flying experience with the WASP as a basis for the issue of rating and instrument cards—in the manner this was done for rated military pilots. Transfers of qualified WASP personnel to available clerical, professional, or technical Civil Service positions were encouraged.[21]

November 20 was the date after which a WASP was allowed to leave "with written approval of her commanding officer stating that her separation would not be a detriment to the war activities at his station."[22] Those who left after November 20, with their CO's permission, were credited with full service and entitled to its benefits. Those who left before then were not.

November 1944, with the deactivation of his women pilots looming, General Robert E. Nowland, who had replaced General Tunner as Ferrying Division commander as of August 1944, wanted a report card on his WASP before they left the service. In a letter dated November 24, 1944, he asked the commanders of the six bases that housed his WASP ferry pilots—the men and their personnel who worked with the women on a daily basis—to assess their contributions.[23]

The women's male peers' consensus was:[24]

1. The women should have Reserve status.
2. Women pilots, as a group, lacked knowledge of the mechanical side of aviation and that was a drawback.
3. Women pilots, as a group, lacked the physical strength to handle larger, heavier twin- and four-engine aircraft under emergency and adverse weather conditions. (The use of hydraulics in aircraft was still in its early stages.[25])
4. Women pilots would best be limited to ferrying pursuit and training type of aircraft.

Number four, of course, embraced the very roles the women had, over the course of two years, filled with success.

Capt. Joe Tracy, who worked with the original WAFS in Wilmington and who continued to work with the women ferry pilots at several bases until their deactivation, gave them a staunch endorsement.

"This brief outline," he wrote, "will not give much of a picture of the splendid work done by the WASPs, but the writer's personal feelings in this matter are that he is extremely sorry to see this splendid organization de-activated at this time and he fears that the loss of these pilots will be keenly felt, especially in the Ferrying Division." [26]

Now the scramble was on to find male pursuit ferry pilots to take over all pursuit deliveries for the Ferrying Division. Pursuit school classes at Brownsville were expanded to take 140 a month rather than the usual 110, but the Ferrying Groups were short sufficient personnel to meet pursuit school quotas.[27]

Ninety P-51Ds sat on the ground at North American in Inglewood, California, the end of November. The Bell Modification Center at Niagara, New York, had eight P-63s backlogged. And Great Falls, Montana, had twenty-seven P-63s awaiting male pilots to fly them on to Alaska.[28]

The WASP at Long Beach sent a radiogram to General Nowland on December 14 offering to serve as ferry pilots on a volunteer dollar-a-year basis "until such time as the present necessity for pilots is alleviated and sufficient pilots are trained to replace us."[29]

Other WASP squadrons followed suit. All offers were declined.

AAF combat organizations offered 161 male pilots, on loan, to the Ferrying Division for thirty days to take care of the P-51s, P-63s, and P-47s waiting to be delivered. The historic details were lost, so exactly where these men were coming from is unknown. The P-51 pilots were to report to the 6th FG in Long Beach and the P-63 pilots to the 3rd FG in Romulus, within a few days after December 20.[30]

Nowland requested an additional fifty pilots from the 6th Air Force to ferry the stacked up P-47s, but his request was refused on December 27.[31]

The WASP ferry pilots had at least one friend in the national press. Captain Gill Robb Wilson, an aviation expert on the staff of the *New York Herald Tribune* (later the founder and editor of *Flying Magazine*), criticized the AAF's waste of flying abilities brought on by the deactivation of the WASP of the Ferrying Division:[32]

> Training men to take their places will require a million dollars, from four to six months' time, and even then will not replace the broad experience which the women have built up on pursuit-type aircraft. I just do not think the United States is rich enough to throw away two hundred expert pilots, regardless of who they are. ...

An editorial in that same issue of the *Herald Tribune* supported Wilson's points.

The tide of public opinion already was turning against women performing in jobs that were not traditional for them. Better said, the tide was skillfully being turned against them by conservative factions who wished to return to the old order.

The whole tenor of the war had changed and the country already was moving on. A return to normal was the idea, whatever normal was. And women flying military airplanes, serving as school principals, and working in munitions factories was not "the norm." We would first have to learn to live with peace and prosperity and our country's new leadership role in the world. Women's rights and abilities would reassert themselves, but it would take a generation to come to fruition.

Both Nancy Love and Jacqueline Cochran were ahead of their time.

That Cochran felt Nancy Love and the command of the FD/ATC had it in for her is borne out in an undated letter to General Arnold. The eight-page document was written on Jacqueline Cochran Ranch, Indio, California, letterhead after the WASP program was disbanded. "Memo for General Arnold —Very personal and confidential." Some excerpts:[33]

> You asked me where I was having trouble in connection with the WASP program and I replied I would prefer to tell you the story after completion of the program. ...

> The trouble first, last and all the time generated in the Air Transport Command. ... Gen. C.R. Smith ... wrote you ... that the operations within ATC were such that each squadron of women would report through a squadron leader to a base Commander and there could be no chief of these various squadrons. How specious this was is disclosed by their own action ... in appointing Nancy Love to the self same post. ... I always considered C.R. Smith the worst offender because he knew better. ... I heard it said that Col. Robert Love helped C.R. Smith get something he needed in Boston in the prewar aviation days and Smith repaid Love by bringing him into the ATC program and by attempting to build up his wife. ...
>
> I can't prove it but am sure that most of the articles against WASP or WASP militarization, which sometimes degenerated into personalities, originated from contacts within ATC. ... All this could have been stopped early in the war if I had been given power. ...
>
> So far as Nancy Love was concerned she was ambitious. ... [Nancy Love] and her close friend were given special transition to fly a B-17 across to Europe and thus was stopped because it was against the policy to use women pilots other than in the United States and Canada. She blamed that on me and her hatred grew to the explosion point where she lost her sound reasoning powers and was against everything I tried to do. ... If I had been empowered to fire her she would either have quit her activities or quit her job. As it was I suffered along rather than have anyone think it might be a personal scrap between two females.

Nancy Love and Betty Gillies were stopped from flying the North Atlantic on order of General Arnold September 5, 1943. The regulation against women flying "overseas" did not become official until AAF Regulation 40-8, as revised, went into effect December 21, 1943.

Cochran concluded with:

> One more word about ATC. A friend of mine who is special consultant to the Under Secretary of War and has a son-in-law in

> the ATC thought, until I cleared his mind ... that ATC was entirely separate from the Army Air Forces. The name Air Transport Command is constantly used in all publicity without any reference to Air Forces and lends support to this apart and independent idea. It should be called "Air Forces Transport Command."

The memo is remarkable. To this reader, it reveals an egocentric mind that took the sum of events of twenty-seven months and twisted them around to fit her own needs and perceptions, making herself appear to be the injured party in a grand scheme against her. Cochran appears to have forgotten that she tried to write Nancy Love out of a job in the proposed changes for the WASP before Regulation 40-8 went into effect in December 1943 (see Chapter 18). And Cochran's grievances resurfaced in her later years, as borne out in her oral histories done in her declining years.[34] To other readers, the memo may send other messages. No doubt Cochran felt wronged and thwarted, and took the hurts seriously to heart.

Nancy Love, not long before her death in October 1976, spoke with then president of the WASP organization, Bee Falk Haydu, 44-7. Nancy told Bee, "the first thing I did when I got home after deactivation was burn my WASP uniform."[35]

Yet another manifestation of taking hurts seriously to heart. We are all human.

The Vagaries of War

As the WASP faced the shut-down and removal from active service, on December 5, 1944, the call went out for German women over eighteen to volunteer for the army or air force.[36]

Chapter 32

Thoughts: Women of the Ferry Command … In Their Own Words

The WASP's oral histories are a rich source of stories about their service and dedication, but also about their personal takes on events, attitudes and inevitabilities. Enjoy!

Betty Huyler Gillies, WAFS—Wilmington[1]

I couldn't see very well. It was still quite dark—a typical morning in Long Beach, with scud clouds and the ceiling 800 to 1,000 feet. But when they said go, I went. Shortly after I was airborne, the airplane burst through the murk into a cloudless blue sky.

The sun's rays hit the silver wings of that P-38 and turned them to gold. I wanted to pull something and stop right there in the air. Below was fog and black stuff; directly in front of me were these two snow-covered peaks[2]—well over ten thousand feet; the sun was coming up through the pass between them; and up above it all was this sleek airplane flying on wings of gold.

RUTH DAILEY HELM, 43-2 —DALLAS[3]

The P-51 is a sweet airplane. It's easy to bring it in over the fence at 120 miles an hour—pull the throttle and set it down. I never had a problem. Just point me to where you want me to go. Later I checked out in the P-38. It was a sweet airplane, too.

I was the Operations Officer in Dallas. Some of the girls under me liked to stay on base and go to the officers' club, so if a good trip came up I could sign myself on it.

Later, there were guys at Dallas who had come back from Europe. They were placed with the Ferry Command and they were miserable—just sitting there. They couldn't go home. It was not that the Ferrying Division had more pilots than it needed—it *had* to keep them. These guys were trained to fly only one aircraft. The Ferry Command was not the best place for them, but it's the only place the military had for them. They sent the WASP home about three months too early because, right after the military deactivated us, they had an awful time getting pursuit ferry pilots. They needed the women pilots.

My own personal feelings? I wasn't angry. I could see where it was better to let us go than the men, because they couldn't let them go. But I thought we got a pretty rough deal. I was checked out on the P-38 and the B-25 because the Ferry Division was trying its best to be able to hire back a certain number of women. At least that was the undercurrent. I don't know if there was anything official about it, but all of a sudden a lot of us got pulled off of P-51s and were flying P-38s and checking out in B-25s, both of which gave us a Class IV rating. Cochran, of course, was against it. When everybody was deactivated, there was no rehiring.

Photo 43. Ruth Dailey, Class 43-2, a Dallas ferry pilot, also flew the P-38.

Photo courtesy of the Iris Cummings Critchell Personal Special Collection

My theory? I think in the back of their minds, the Ferrying Division figured that there would be a cadre of, say, fifty of these women they might have been able to keep.

Gayle Bevis Ewing Reed, 43-5 – Dallas[4]

When you flew your first pursuit, it had one seat. They'd pat you on the shoulder and say, "I know you can do it."

I drew a P-39 first. I got in the airplane. It didn't sound too different. I checked my mags, called the tower for clearance. They gave it to me right away. But I was not in that big of a hurry. I pulled out on the runway. ... That airplane jumped off the ground and I never felt such power! I felt like I was attached to this big fan—the propeller. I was over the Gulf of Mexico before I ever remembered to put my wheels up. I lowered the speed, made my level turn, my climbing turn, and then went out in the practice area.

I flew around making gentle turns. Later on, I wrapped it up a little more. There were some clouds that I was supposed to avoid. But they still had holes in them. So, I flew down to them, saw what it was on my altimeter, climbed 1,000 feet above them and practiced landing on the clouds.

Now the P-51. Oh God! It was the high spot of my life. That Mustang that Pinkard [her instructor at Sweetwater] had told me about? I had my moment. Not everybody gets it.

Jo Myers Wheelis, 43-5 —Romulus[5]

One memorable flight, after taking a P-63 to Boise, Idaho, I caught a cargo plane out of Salt Lake City back to Romulus. The mountains are pretty high east of Salt Lake City. You had to circle until you got up high enough to get over the mountain. When we got up there, right on top of the clouds, the moon was shining and it was snowing. And those peaks were sticking up through the clouds. I'll never forget what that looked like.

One time I was going to Washington D.C. to see my sister, so I bummed a ride down to Wilmington. I still needed to get to D.C. Nancy Love was there in her plane—a single-seater, twin-engine, but a rear compartment for a gunner.[6] She asked if I wanted to go with her. I said, "Sure." She put me in the gunner's seat, flew me down to Washington, and drove me out to my sister's house. That was a good association I have never forgotten.

Photo 44. Jo Myers, Class 43-5, Romulus ferry pilot.

Courtesy WASP Archive, Texas Woman's University, Denton, TX.

We had a lot of privileges at Romulus. We weren't officers, we were civilian pilots, but we could go to the Officers' Club. We were allowed officers' privileges, so we went there and ate whenever we wanted to. But I saw the end coming. It wasn't being run like it had. The attitudes weren't real good. There wasn't any future there. I left in October.

Just saying I was in the WASP doesn't mean anything. It's remembering what I did and how it all turned out, what I experienced and the education I got. That's all a part of it. You can't just say being in the WASP meant something. It meant all those things. I have profited from it many times more than I even realized.

Sylvia Dahmes Clayton, 43-5 —Romulus[7]

When we were on base at Romulus, if we didn't have another plane assigned to us to go pick up somewhere, we were sent to the flight line where we checked out on other airplanes. That's where I got to check out on the C-47. We flew copilot.

Betty Scantland (43-6) and I were assigned to pick up a C-47 and take it to Montreal for modification. Betty was shorter than I am. When we landed, there were no linemen to put the control locks on the plane or put the chocks under the wheels. Betty and I had to get out and do that.

When we got inside, the officer said he needed for the pilot of that C-47 to sign in and close the flight plan. I said, "I'll do it." He said, "No you won't. I want the pilot."

I said, "I am the pilot."

He looked at me.

Betty said, "She is the pilot. She's the pilot of that plane."

He looked at me again and said, "Okay."

He just had not seen a woman flying one of those planes.

And, yes, it was true—at least at Romulus—that the wives of the male pilots asked that the WASP not be sent out as copilots with their husbands. We always felt, well, they knew their husbands.

Anna Flynn Monkiewicz, 43-6 —Romulus / Wilmington[8]

Not long after returning from pursuit school in August 1944, I was transferred to New Castle. I liked New Castle. There, I got to fly with the remainder of the WAFS—they were great gals. We spent much of our time on detached service out on Long Island where we flew P-47s. In fact, I got to deliver the last P-47 for the girls on December 19. There was one more to be delivered and we drew matchsticks. I got the good one.

Because of making that last delivery, I didn't get back to Wilmington in time for the big party thrown the night of December 19; consequently, I'm not in the last day photo.

Deactivation? We felt like the rug had been yanked out from under us. We hadn't finished our job. The operations people said, "There's plenty of work here to be done, why are you sending them home?"

I blame Jackie. Maybe that is unwarranted ... I know she was trying to get the WASP into the military as a separate unit. She did not want to work under Hobby, but there weren't enough of us to establish a new unit. We could have stayed Civil Service and that's what I would have wanted. I thought, why don't you just leave us alone? Let the war wind down and then send us home. Maybe that's what saved my life. Who knows?

[And from her book *My Piece of the Sky*:] The hole I'd left at home ... two years earlier ... had quickly filled in, as when you drop a stone in the water. There didn't seem to be room for me any more. ... I didn't quite fit in. I didn't know what to do with myself. I just wasn't able to settle down. ... In March of 1945 I got a phone call from Rusty McLean (43-7), who was in Ohio with two other WASP. "Come on out, Flynn.

You can get your instructor's rating and teach." ... And so began another chapter, another piece of my sky.[9]

Dorothy Kocher Olsen, 43-4 — Long Beach[10]

Because I had those A-24 flights with Iris Cummings back in the fall of '43, when I got to pursuit school, I was the first one to check out. My first flight was in a P-47. It comes down like a rock, it's so heavy, and I greased it in. I'm good at that sort of thing. My instructor was so proud of me, because all the instructors were on the flight line watching. I was the first woman in the class to come in in a fighter at pursuit school.

Well, I just loved the P-51 from day one. I checked out in the P-38 and loved it too. The P-51s and the P-38s almost always went to Newark. And then from there they would take us, in the military airline over to Niagara Falls, where we'd pick up a P-39 or P-63. From there we would go to Great Falls, Montana. But they wouldn't let us women take them on to Alaska.

The P-51 was a lot harder to fly than the 38. The P-38's like driving a car. It's the easiest plane to fly as long as everything's working right. Lose an engine in a P-38 on takeoff, you're in trouble. I think a P-51 you could handle, but not a P-38.

I loved every single minute of it!

[The day we were disbanded] was the saddest day of my life because we didn't know how near over the war was. We had no idea about the atomic bomb. My brother was still on a submarine out there being bombed, you know, depth bombed. He had some really bad times.

It felt like being fired. And I have always blamed Jacqueline Cochran. Nobody else does, but I do. I think she didn't get her way, so if she can't have her way we all go home. I think that was the cruelest, meanest thing that anybody has ever done, to be that close to the end and send us home. I'll never forgive them for that. I came home with my tail between my legs.

When I left the base, the whole field was full of P-51s waiting to be delivered and nobody to deliver them. We said we'd fly without pay. And our base commander wanted to keep us. But Jacqueline Cochran? She can't have her way, everybody goes home.

***

Photo 45. Ginny Hill, Class 43-4, and Barbara Jane Erickson, WAFS, Long Beach ferry pilots.

Photo from Sarah Byrn Rickman's personal collection
This was the first of several B-17s that Erickson delivered. Hill was her copilot.

Ginny Hill Wood (43-4) – Long Beach[11]

I remember the last flight I took. It was a P-51. I remember that all those guys in my ground crew lined up and saluted me. They didn't have to do that. I didn't care about that military stuff. ... But I thought that was pretty neat what my ground crew did and I felt very good about that.

When I got out, I was given a chance to be a first lieutenant at a desk job. I turned it down. I wanted to keep flying. I was a pilot, for goodness sake! And I didn't want any more war stuff. I was, by instinct, a Quaker. I think it just bothered me some. I guess I rationalized well, you know, what's the difference between flying airplanes and making them? Just be grateful you got good eyesight so you get to fly. You have a privilege a lot of people don't.

***

Ruth Adams, 44-2 —Wilmington[12]

When we were deactivated I wasn't of two minds. Those goddamn planes could kill me or I could kill myself by doing something dumb. My main feeling, about 60 percent of me, was glad that I had made it and was still alive.

Suddenly, we were dropped. We got letters that, in effect, said that the government appreciated our services, our contributions to the war effort, but now it was goodbye—rather like a pat on the head, "you were cute little girls." Women were allowed to be schoolteachers, stenographers, or nurses, but wanting to be pilots was not acceptable. I sensed that four out of five men resented us because we took the glory out of flying. They couldn't swagger as much because we could fly, too.[13]

The WASP experience gave me self-confidence.

***

Jary Johnson McKay, 43-2 —Romulus[14]

Mostly we flew P-39s out of Niagara Falls. After pursuit training, I had a flight leader for a couple of flights, but usually we were totally on our own. We would be sent to pick up and deliver those planes in a bunch. But we were totally on our own to get to our destination. We were supposed to be off at sunrise but nobody ever was. We had certain fields to land at for fuel. We couldn't just land at any old airport. And we always had to stop at night. The military would have some kind of vehicle to take us to a hotel. We always doubled up, because we were on per diem and we wanted to save our money. It was up to us to make the per diem stretch.

I've heard that some WASP thought the male pilots resented them. There may have been some who resented us, but the men who were my friends accepted me or they wouldn't have been my friends. In fact, I was dating one of them for quite a while. I think we steered clear of those other guys. Sure, it was a common thing for some of those men to make a practice of going in town and picking up women. I would hear them bragging the next morning. Now, that's the kind of men that may have resented women pilots. Those guys weren't my friends.

We were all out ferrying—all over the country—when they called us back in December to be deactivated. I was in Niagara Falls. I had been assigned a P-39 and it was snowing and there were drifts on the airport so we couldn't take off and we were waiting. Finally I was cleared to leave, but I got a telegram to report back to Romulus. As I remember, it took me by surprise. I don't remember whether I suspected that the program was coming to an end or not. I had been away from the base so much, I just didn't hear.

The publicity emphasizes "oh wasn't this awful, that they just sent you home! They didn't give you any benefits!" I don't feel that way. I feel so lucky to have had the experience. After all these years I hate to see all this bitterness. It takes away from it. Some of the women are very bitter about the men. But most of us weren't that way.

When people I know say, "wasn't it awful the way they treated you!"—and that's the first thing they say—I remember the great thing was that we got to fly.

***

Iris Cummings Critchell, 43-2 —Long Beach[15]

Of course, we knew the end was coming and prepared for it. Some of us were interested in continuing in aviation, professionally, after the war and we knew it would be highly desirable to have a CAA (civilian) Instrument Rating. The bases offered the WASP pilots who were interested the opportunity to practice all the variations on the instrument approaches in the landing simulator—and lots of Link time. They also provided the gas budget and the airplane sitting on the ramp for them to fly. Katy Loft (43-4) and I took advantage of the opportunity. One of us would fly observer and the other would fly under the hood. Then we would trade off.

When we were ready, before we left December 20, we arranged for the flight test for the CAA rating. Then they offered us night qualifications in the twins and, of course, we jumped at the chance. Near the end, I did night qualifying in the C-47, A-20, P-38, and P-61 by flying after sundown for an hour or two. I left the Ferry Command current in instruments and night flying as well and with a coveted CAA Instrument Rating.

In November, I was flying P-61s across the country to Newark. Then, for some reason, I got put back on P-38s. The first week of December, I was doing a round robin every single day—P-38 movements up through December 10. Then I finished up with mostly P-61 deliveries. December 13, I flew a P-61 to Sacramento, and an A-20 from Sacramento back to Long Beach. Sacramento was a mod center. We were being kept close to base, so we wouldn't be caught out some place when the December 20 deadline came.

December 15, another P-61 to Sacramento, and on December 17, one more P-61 to Sacramento, and an A-20H from Sacramento back to Long Beach—my very last flight with the Ferry Command.

***

Jean Landis, 43-4 —Long Beach[16]

We had no rank, no life insurance, no medical insurance. We were volunteers, civilians. If we didn't like it, we could go home. No one I knew quit.

I flew all the single-engine pursuits, but the P-51 was my favorite—its looks—and that engine makes a beautiful sound. You can recognize a P-51 by the sound. I still get chills when I hear one. Strapped in, I felt like part of the plane, like a bird and those were my wings.

On my final WASP flight, I approached Newark, New Jersey, in my very last P-51. Before landing, I circled the Statue of Liberty. Today you can't do that. You'd be shot down. But I circled and waggled my wings and I wept. And that was it.

Jacqueline Cochran wanted to have her own command and she wouldn't accept our going into the WAC. So, though we were needed, we were deactivated. No one said thanks. Just, "Turn in your gear and go home to mother." Everyone I knew would have gone into combat if we had been allowed.

I went back to take up my teaching job that I had left in February 1943. Because I was going into the military, my job was to be held for me. But we were not militarized. Since I wasn't military, they gave my job to someone else. I ferried surplus airplanes for a bit, but after flying the Rolls Royce of the sky, the P-51, ferrying beat-up trainers wasn't very appealing. Finally, I quit aviation altogether.

I did return to teaching—to my great joy and satisfaction. I earned a Masters in health, physical education, and recreation from Wellesley, and taught at the college level in four different states for a total of thirty-four years. I finished my career back at my alma mater, by then known as San Diego State University, and retired in 1980. I've traveled abroad and I've traveled the USA in my RV. I found a little piece of heaven in northern Idaho, thirty miles from the Canadian border, and lived there for thirty years. Now I'm home again in El Cajon.

Every experience in life changes us. I would do it all over again in a heartbeat.

Photo 46. Long Beach ferry pilot Jean Landis, Class 43-4, with her favorite aircraft, the P-51 Mustang.

Photo from Jean Landis' personal collection.

Afterword

> Historians James Lea Cate and Wesley Frank Craven, editors, *The Army Air Forces in World War II, Vol. VI: Men and Planes:* "by the time [the women] were ready to replace a substantial number of men, in keeping with the original purpose of their organization, victories overseas brought a reduction in military requirements for pilots."[1]

The morning of December 20, 1944, as she climbed into her car and prepared to head home to Seattle, Barbara Erickson — C.O. of the WASP ferrying squadron at Long Beach — noted "sixty-six new P-51s sitting beside the runway waiting to be delivered. And they wouldn't get delivered that day or the next because the qualified women pilots who had been ferrying P-51s to Newark—and who were still so badly needed—had been dismissed and sent home."[2]

A similar dilemma—new P-47s and P-63s, as well as P-51s—existed at Farmingdale, Buffalo, Niagara, Dallas, and Evansville.

Photo 47. Barbara Jane Erickson, WAFS, Squadron Commander at Long Beach, with one of the P-38s she ferried.

Photo courtesy of the Iris Cummings Critchell Personal Special Collection

As the WASP headed home on December 20, and in the days after, the newspapers were full of reports from the Battle of the Bulge. No one paid any attention to the release from duty of 905 women who had been flying U.S. aircraft for the Army Air Forces in an effort to help win that very war.

***

When the war was over, most WASP went home, married, had children, settled in to raise them, and went on with their lives. Some went to work. Some went back to school. A few managed to stay in aviation. Notably, they didn't talk about their wartime service because they quickly learned that people did not believe them when they did. The AAF sealed their records. Life went on.

The WASP were middle-aged when a news story ran in the mid-1970s lauding a group of U.S Navy women as the first American women to fly military aircraft.

"We were the first," the WASP declared, collectively. A renewed effort erupted to win recognition by gaining their militarization. Accomplishing that required a huge effort on the part of the WASP themselves, and from their ardent supporters: Senator Barry Goldwater, who had served side-by-side with them in the Ferry Command; Col. Bruce Arnold, son of the late Gen. "Hap" Arnold, who dedicated himself to the realization of his father's goal, WASP militarization; the female Assistant Secretary of the Air Force, Antonia Handler Chayes; and all four women in Congress at that time—Representatives Lindy Boggs (D-Louisiana), Margaret Heckler, (R-Massachusetts), and Elizabeth Holtzman (D-New York), and Senator Margaret Chase Smith (R) of Maine.[3] WASP militarization was granted in 1977.

The long-sealed WASP records were released in the late 1970s and books about the women began to appear.

Sarah Byrn Rickman, December 2015

Endnotes

Dedication

1. Debra Winegarten, *Katherine Stinson: The Flying Schoolgirl* (Austin, TX: Eakin Press, a division of Sunbelt Media, 2000), 79. Pioneer aviatrix Marjorie Stinson, born 1896, trained eighty-three pilots for World War I. Marjorie was Katherine's younger sister. Ruth Law, born 1887, campaigned, though unsuccessfully, for women to fly for the Army in that war: http://en.wikipedia.org/wiki/Ruth_Bancroft_Law Accessed 8-17-2013. Charles E. Planck, *Women With Wings* (New York: Harper and Brothers, 1942), 80-81. "[Ruth Law] was authorized to wear a military uniform and spent some time posing for recruiting posters. She was also authorized to teach military flyers, but her Liberty Bond flights and recruiting assignments kept her too busy to do much instructing."

Foreword

1. Eleanor Roosevelt, "My Day," 1 September 1942, last accessed 10 September 2015, https://www.gwu.edu/~erpapers/myday/displaydoc.cfm?_y=1942&_f=md056279

2. Letter, Brigadier General W.H. Tunner, Commanding General, India/China Division, ATC to CG Ferrying Division ATC, "Commendation," 16 October 1944, quoted in "Statement by Dora Dougherty Strother before the Committee on Veterans' Affairs, House of Representatives, 95th Congress 1st Session, in support of HR 3321 Introduced by Congresswoman Lindy Biggs, 20 September 1977," in Congress, House, Select Subcommittee of the Committee on Veterans' Affairs: *Hearing before a Select Subcommittee of the Committee on Veterans' Affairs*, 95th Cong., 1st sess., September 20, 1977, p. 312.

3. Address by General H. H. Arnold, Commanding General, Army Air Forces to Last W.A.S.P. Graduating Class, 7 December 1944, last accessed 10 September 2015, http://wingsacrossamerica.us/records_all/press_archive/arnold_press1.pdf.

4. Cornelia Fort, "At the Twilight's Last Gleaming," *Women's Home Companion* (July 1943), p. 19, last accessed 10 September 2015, http://www.wingsacrossamerica.us/records_all/wasp_articles/twilight.pdf

5. Deborah G. Douglas, *American Women and Flight Since 1940* (Lexington: University Press of Kentucky, 2004).

Acknowledgments

1. National Aviation Hall of Fame Combs Gates Award website: http://www.nationalaviation.org/combs-gates/

Part I

1. Dominick A. Pisano, *To Fill the Sky with Pilots* (Urbana: University of Illinois Press, 1993), 3.

2. Patricia Strickland, *The Putt-Putt Air Force: The Story of the Civilian Pilot Training Program and the War Training Service (1939–1944)* (Washington D.C., Department of Transportation, Federal Aviation Administration, Aviation Education Staff, GA-20-84, ca. 1970), iii.

3. Pisano, *To Fill the Sky,* 77.

Chapter 1: The Beginnings

1. William "Billy" Mitchell (December 29, 1879–February 19, 1936) is the United States Army general who is regarded as the father of the U.S. Air Force. He advocated air power for future wars, arguing that bombers could sink battleships. He antagonized many people and in 1925, he was

court-martialed. http://en.wikipedia.org/wiki/Billy_Mitchell_(general) Accessed 3-4-2012

2. William H. Tunner and Booton Herndon, *Over the Hump: The Story of General William H. Tunner, the Man Who Moved Anything Anywhere, Anytime* (New York: Duell, Sloan and Pearce, 1964), 6.

3. Tunner, *Over the Hump,* 13–14.

4. The Ninety-Nines are licensed women pilots. The group was founded on Nov. 2, 1929, in Valley Stream, Long Island, by Amelia Earhart and 98 other pioneer women flyers. Earhart was their first president. The name 99s comes from their charter membership. http://www.ninety-nines.org

5. Doris L. Rich, *Jackie Cochran: Pilot in the Fastest Lane* (Gainesville: University Press of Florida, 2007), 99–100. Her source is the *Los Angeles Times,* May 11, 1936.

6. Rich, 99.

7. "Women Pilots in the AAF, 1941–1944," *Army Air Forces Historical Studies,* no. 55 (March 1946): 2–3. Copies held by the WASP Archives, Texas Woman's University Library, Denton, Texas, the Jacqueline Cochran collection, Dwight D. Eisenhower Library, Abilene, Kansas, and the author. Referred to hereafter as #55, "Women Pilots AAF."

8. Letter from Nancy Love to Col. Robert Olds, May 21, 1940, Nancy Harkness Love private collection (NHL collection), held by her daughter, Margaret Love.

9. #55, "Women Pilots AAF," 3-4.

10. Lt. Col. Oliver LaFarge (Air Force Reserve), *The Eagle in the Egg* (Boston: Houghton Mifflin, Riverside Press, 1949), 9.

11. Lend-Lease (Public Law 77-11) was the program under which the United States of America supplied the United Kingdom, the Soviet Union, China, Free France, and other Allied nations with materiel between 1941 and 1945. http://en.wikipedia.org/wiki/Lend-Lease — accessed 3-3-2012.

12. Tunner, *Over the Hump*, 18.

13. Tunner, *Over the Hump*, 19.

14. Tunner, *Over the Hump*, 20.

15. Tunner, *Over the Hump*, 20–21.

16. "Women Pilots of the Air Transport Command," a history compiled by Lt. Col. Oliver LaFarge, and the accompanying Appendix II, Memorandum for the Chief of the Air Corps, dated August 5, 1941, pp. 30-32. Texas Woman's University Library, Woman's Collection, WASP Archives. Copies of both in the author's hands.

17. LaFarge, *The Eagle in the Egg*, 1.

18. LaFarge, *The Eagle*, 2-3.

19. LaFarge, *The Eagle*, 4.

20. Iris Cummings, WASP Class 43-2 and aviation historian who holds emeritus professor status with Harvey Mudd College, Claremont, CA. From contents of an email sent to the author July 15, 2013, and titled "Brief History of factors in the Origins of Ferry Command procedures." They are taken from the manuscript of her memoir—working title "76 Years With the Spirit of Flight." There will be more of these referenced throughout the manuscript.

21. Tunner, *Over the Hump*, 25.

22. *World War II, Day by Day* (London: Dorling Kindersley, 2001), 262. Also *American Experience, FDR*, DVD Part IV, "The Juggler," 1995.

23. Letter from Jacqueline Cochran to General Arnold marked "Urgent and Important." Photocopy undated, but military document says January 18, 1942. Cochran collection, Eisenhower Library.

24. #55, "Women Pilots AAF," 11.

25. Email to author, 11-17-2015, from Robert B. Patterson, grandson of Col. Robert H. Baker.

26. Col. Onas P. Matz USAFR (Ret.), *History of the 2nd Ferrying Group.* Sponsored by the Wilmington Warrior Association (Seattle, WA: Modet Enterprises, 1993), 6-7.

27. Mardo Crane, "The Women with Silver Wings," Part One, *The 99 News,* Special Issue 1978, clipping found in NHL collection.

28. https://en.wikipedia.org/wiki/Harold_L._George; During World War II, George led the Air Transport Command, taking it from 130 obsolescent aircraft to 3,000 modern transports, operated by 300,000 airmen. Accessed 11-17-2015.

29. H.H. Arnold, *Global Mission* (Blue Ridge Summit, PA: TAB Books, Military Classics Series, 1989, by arrangement with Harper & Row Publications, Inc., 1949), 294.

30. Robert Serling, *Eagle: The Story of American Airlines* (New York: St. Martin's/Marek, 1985), 46, 161-162; and James M. Mangan, *To The Four Winds: A History of the Flight Operations of American Airlines Personnel for the Air Transport Command, 1942-1945, including Project 7A,* (Paducah, KY: Turner Publishing Company, 1990), 7. Also Arnold, *Global Mission,* 295.

31. Cyrus Rowlett (C.R.) Smith resigned as president and director of American Airlines to enter the Army in April 1942. Source: Airlift/Tanker Association Hall of Fame: www.atalink.org/hallfame/c.r.smith.html

32. Arnold, *Global Mission,* 295.

33. Serling, *Eagle,* 162-163.

34. *Baltimore Sun,* Apr 19, 1942, copied to the author by Robert B. Patterson, Colonel Baker's grandson.

35. Lt. Col. Oliver LaFarge, "History of the Air Transport Command, Women Pilots in the Air Transport Command," prepared by the Historical Branch, Intelligence and Security Division, Headquarters, Air Transport Command. Lt. Col. LaFarge is official historian for the Air Transport Command, 13. This is the accepted history on the women ferry pilots

of the ATC. (See footnote #32 below for additional explanation.) From the private collection of Mrs. Ann Hamilton Tunner, widow of General William H. Tunner. A copy also is in the hands of the author. Hereafter, LaFarge, "Women Pilots ATC."

36. LaFarge, "Women Pilots ATC," 13-14.

37. LaFarge, 14.

38. Email to author, Nov. 17, 2015, from Robert B. Patterson, Col Baker's grandson.

39. LaFarge, footnote #18 and the bottom of page 13.

CHAPTER 2: A COUP AT THE WATER COOLER

1. Tunner, *Over the Hump,* 34–35. General Tunner's written recollection of the conversation as quoted in his book.

2. Sarah Byrn Rickman, *Nancy Love and the WASP Ferry Pilots of World War II* (Denton: University of North Texas Press, 2008), 36–39, 54–58, 70.

3. Memorandum: Colonel Tunner to General Harold George dated June 18, 1942. Subject: Use of Women Pilots in Ferrying of Aircraft, with copies to Col. C.R. Smith and Lt. Col. Robert H. Baker. Source: Appendix II, LaFarge, "Women Pilots ATC"; also #55, "Women Pilots AAF," 12–14.

4. Betty Gillies, "The When, Why, Who, Where of the WAFS" presented at the Southern California WASP Meeting, Laguna Hills, CA, June 14, 1987. The Betty Gillies Collection, the International Women's Air and Space Museum, Cleveland, OH.

5. Memo: June 18, Tunner to George, Appendix II, 34-35; #55, "Women Pilots AAF," 13, fn 28; and LaFarge, "Women Pilots ATC," 14, fn 21.

6. Memo: George to Arnold, July 14, Subject: WAFTs (Women's Auxiliary Ferry Troop), the first name given the group, from LaFarge, "Women Pilots ATC," Appendix II, 37.

7. Memo: George to Arnold, July 14, Appendix II, 9; LaFarge, "Women Pilots ATC," 14–15, 18.

8. LaFarge, "Women Pilots ATC," 15–16.

9. LaFarge, "Women Pilots ATC," 16–17; Marx, "Women Pilots Ferrying Division," 17–23.

10. Marx, "Women Pilots Ferrying Division," 26–27.

11. LaFarge, "Women Pilots ATC," 18.

12. Routing and Record Sheet, Headquarters AAF, Subject: WAFTs, from George to Arnold July 18, 1942, Appendix II, 38-39

13. Routing/Record Sheet, Subject: WAFTs, Arnold to George, July 20, 1942, Appendix II; LaFarge, "Women Pilots ATC," 20–21.

14. Delphine Bohn, unpublished WAFS memoir "Catch a Shooting Star," Chapter 4, 16–17. Copies in hands of the author and at the WASP Archives, TWU.

15. LaFarge, "Women Pilots ATC," 19.

16. Eleanor Roosevelt, "My Day," syndicated newspaper column dated September 1, 1942, copy in TWU.

17. Memo dated Sept. 3, 1942, from General George to General Arnold, Subject: Employment of Women Pilots.

18. Routing dated Sept. 5, 1942, "General Arnold directs that immediate action be taken on this project and expects that the initiation of the recruiting of these women will take place within 24 hours." Signed Brig. Gen. L.S. Kuter, Appendix II, 42; and LaFarge, "Women Pilots ATC," 22–23.

19. #55, "Women Pilots AAF," 16, fn 35.

20. LaFarge, "Women Pilots ATC," 23, 29.

21. LaFarge, "Women Pilots ATC," 23, and fn 35.

22. LaFarge, "Women Pilots ATC," 29.

23. LaFarge, "Women Pilots ATC," 28.

24. Jacqueline Cochran and Maryann Bucknam Brinley, *Jackie Cochran, An Autobiography* (New York: Bantam Books, 1987), 197. Similar references are found in Rich, *Pilot in the Fastest Lane,* 113, and also in *The Stars at Noon,* by Jacqueline Cochran (Boston: Little, Brown and Company, 1954), 117.

25. Arnold, *Global Mission,* 311.

26. #55, "Women Pilots AAF," 17

27. Deborah G. Douglas, "WASPs of War," *Aviation Heritage,* January 1999.

28. Memo from Jacqueline Cochran to General Arnold, September 11, 1942; subject: Use of Women Pilots; Cochran / Eisenhower.

29. Memo, September 11, 1942.

30. #55, "Women Pilots AAF," 17, fn 41, Cochran, "Comments."

31. LaFarge, "History of the Air Transport Command," 23, fn 35. Source, Lt. Col. Oliver LaFarge's interview with Nancy Love.

32. Memo from Major General H.L. George to General Arnold (copy to General George E. Stratemeyer), September 12, 1942, Appendix II, 44-47; Subject: Training of Women Pilots; Cochran/Eisenhower; TWU WASP Archives.

33. Memo, Arnold to Stratemeyer, September 13, 1942.

34. Routing and Record Sheet: Memo, September 15, 1942, Smith to Arnold, Appendix II, 48, LaFarge, "History of the Women Pilots of the ATC," held at TWU, WASP Archives.

35. Women Pilots with the AAF, 21-22.

36. Memo from General George to General Arnold, September 12, 1942; Cochran/Eisenhower.

37. Memo from Captain James I. Teague to Colonel William H. Tunner, September 22, 1942, NHL private collection. Also, Appendix II.

38. #55, "Women Pilots AAF," 21, fn 47.

39. Arnold, *Global Mission*, 356–357.

40. Undated memo from Jacqueline Cochran to General Arnold marked "Very personal and confidential." Cochran / Eisenhower.

41. Rowland Carter, "The Ladies Join the Air Forces," *Flying Magazine,* December 1942. Copy found in Dunbar Library, Wright State University, Dayton, OH. Author has a copy.

42. October 2, 1942, letter from John S. Wynne to Nancy Love, Nancy Love Collection, International Women's Air and Space Museum.

Chapter 3: Recruiting the WAFS

1. LaFarge, "Women Pilots with the ATC," Appendix II, 59, TWU. Author has a copy. Memo dated September 21, 1942, from Col. Robert H. Baker, Command, 2nd Ferrying Group, Ferrying Division, Air Transport Command, Wilmington, Delaware: Subject—General Regulations re: the Women's Auxiliary Ferrying Squadron.

2. Jan Churchill, *From Delaware to Everywhere: New Castle Army Air Base, New Castle County Airport* (Dover, DE: Dover Litho Printing Co., 2007), 39–40.

3. Churchill, *From Delaware to Everywhere*, 40.

4. A liaison aircraft is a small, usually unarmed aircraft primarily used by military forces for artillery observation or transporting commanders and messages. The concept was developed before World War II and included battlefield reconnaissance, air ambulance, column control, light cargo delivery, and similar duties. Liaison aircraft were able to operate from small, unimproved fields under primitive conditions: http://en.wikipedia.org/wiki/Liaison_aircraft, accessed 12-6-2014.

5. Adela Riek Scharr, *Sisters in the Sky,* vol. 1 (St. Louis, MO: The Patrice Press, 1986), 203.

6. Helen Richards, Air Transport Command Flight Records, in the author's possession.

7. Scharr, *Sisters,* vol. 1, 214-215.

8. Sarah Byrn Rickman, *Nancy Batson Crews: Alabama's First Lady of Flight* (Tuscaloosa: The University of Alabama Press, 2009), 34. From author interview with Crews, June 2000.

9. Nancy Love's logbooks, NHL private collection. Texas Woman's University, memorandum. Love Field is NOT named for Nancy and/or Bob Love. It is named for 1st Lt. Moss L. Love, of Wright County, South Carolina, the tenth Army officer to lose his life in an airplane accident, September 4, 1913.

10. Phyllis Burchfield Fulton, WAFS, telephone interview December 2005 with author Sarah Byrn Rickman.

11. The story of this flight is taken from Teresa James's diary, also from author interviews with Nancy Batson, Florene Miller, and Phyllis Burchfield, and from Delphine Bohn's comments in her unpublished memoir, "Catch a Shooting Star."

12. AAF historian Lt. Col. Oliver LaFarge, "Notes on Interview with Mrs. Nancy H. Love, October 3, 1945," Nancy Love collection, TWU.

13. Sarah Byrn Rickman, *The Originals: The Women's Auxiliary Ferrying Squadron of World War II* (Sarasota, FL: Disc-Us Books, Inc.), 378–379.

14. "WAFS to Fly in Gray-Green Slacks, Jackets," September 16, 1942, article from *Times Herald,* clipping found in NHL private collection. Author has a copy.

15. Bohn, "Catch a Shooting Star," Chapter 5, 9.

16. Marx, "Women FD," 94, 97.

17. Marx, "Women FD," referencing a letter from the Ferrying Division to War Department Bureau of Public Relations, March 29, 1943, subject: "Ghost-writing on WAFS," 101–102.

18. Jeanne Holm (Major General, US Air Force), *Women in the Military. An Unfinished Revolution,* rev. ed. (Novato, CA: Presidio Press, 1992), 22.

Part II

1. *World War II Day by Day,* 338. American women answer industry's call.

Chapter 4: The Women's Flying Training Detachment

1. Lettice Curtis, *The Forgotten Pilots: A Story of the Air Transport Auxiliary, 1939–1945,* 4th ed. (1971; repr. Cheltenham, UK: Westward Digital Limited, 1998), 1, 7–9, 12–21. This is the book to begin with in learning about the women of the ATA, both American, British, and others.

2. Curtis, *The Forgotten Pilots,* 141–147.

3. Compensated by a token salary usually for government service according to the Merriam Webster Dictionary.

4. Women Pilots with the AAF, 23.

5. Women Pilots with the AAF, 25.

6. http://what-when-how.com/women-and-war/united-states-womens-flying-training-detachment-wftd-auxiliary-corps/ — accessed August 27, 2011.

7. Marianne Verges, *On Silver Wings* (New York: Ballantine Books, 1991), 68.

8. Leoti Deaton, Chief Establishment Officer, WASP, an oral history, interviewed by Doris Tanner (44-4) March 20, 1982.

9. Deaton Oral History, Tanner.

10. Leoti Deaton, Chief Establishment Officer, WASP, an oral history, interviewed by Ziggy Hunter, March 18, 1975. Edited by Dawn Letson and Nancy Durr, TWU.

11. Deaton oral history, Tanner.

12. Kay Gott, *Women's Flying Training Detachment, Women Pilots Trained at Houston Texas: A Tribute to Lt. Alfred Fleishman*, 1992, 2.

13. Deaton oral history, Tanner.

Chapter 5: Class 43-1 — the "Guinea Pigs"

1. Byrd Howell Granger, *On Final Approach: The Women Airforce Service Pilots of WW II* (Scottsdale, AZ: Falconer Publishing Company, 1991), 71.

2. Jane Sincell Straughan, WASP Class 43-1, an oral history done by Nancy Durr, the Woman's Collection, Texas Woman's University (TWU), October 4, 1996. Information throughout this chapter and later is taken from that interview, on file in the WASP Archives, TWU, Denton, TX.

3. The ERCO Ercoupe is a low-wing monoplane aircraft that was designed and built in the United States. It was first manufactured by the Engineering and Research Corporation (ERCO) shortly before World War II: online at http://en.wikipedia.org/wiki/Ercoupe, accessed 5-27-2011.

4. Mary Lou Colbert Neale, WASP Class 43-1, an oral history interview done by Dawn Letson, October 4, 2002, for WASP Archives, TWU. Information throughout this chapter is taken from that interview.

5. Col. Oveta Culp Hobby served as Director of the Women's Army Auxiliary Corps (WAAC) later the Women's Army Corps (WAC), which was created to fill gaps left by a shortage of men. Online at http://en.wikipedia.org/wiki/Oveta_Hobby, accessed 5-27-2011.

6. Betty Tackaberry Blake, WASP Class 43-1, an oral history interview by Maura Mackowski, August 6, 2003, WASP Archives, TWU. Information throughout this chapter is taken from that interview. Also, an oral history with Betty Blake, July 14, 2011, interviewer Sarah Byrn Rickman, WASP Archives TWU.

7. Geri Lamphere Nyman, WASP Class 43-1, an oral history interview by Dawn Letson, October 4, 1996, TWU. Information throughout this chapter is taken from that interview.

8. A person who fulfills the necessary requirements is issued a mechanic certificate with either an airframe or power plant rating, or both. It is these ratings which together account for the common practice of referring to mechanics as "A&P's." Until 1952, instead of the power plant rating, an Engine rating was issued, so the abbreviation "A&E" may appear in older documents. http:/en.wikipedia.org/wiki/Aircraft_mechanic. Accessed 12-31-2011.

9. Geri Nyman, "Bean Salads and Bamboo Bombers," Roadrunner Extra, February 2011, a newsletter for the residents by the residents of Beatitudes Campus, Scottsdale AZ. Reprinted in the *WASP News,* Summer 2011 issue, TWU. Courtesy Geri Nyman.

10. Geri Nyman, "Bean Salads and Bamboo Bombers."

11. Celeste Graves, *A View From the Doghouse of the 319th AAFWFTD* (Bloomington, IN: AuthorHouse, 2004), 242. Author's note: W.W. "Bill" Lowrey, Jr., the instructor for Aviation Enterprises who hired Helen Duffy, also instructed the women pilots and made the move to Sweetwater in 1943 where he was appointed Advanced Group Commander of the women's training program, see page 257.

12. #55, "Women Pilots with the AAF, 1941–1944," Chapter II, 27. This document was declassified in 1975. I have copies from both the Eisenhower Library and the WASP Archives at Texas Woman's University.

13. Marx, 189–191; General George, letter dated February 4, 1943.

14. Gen. C.R. Smith to Tunner, March 19, 1943, by command of General George following directive received from General Stratemeyer, Subject: Women Pilots. LaFarge, "Women Pilots of the ATC," Appendix II, 63. TWU WASP Archives.

15. Two sources here: Gott, *A Tribute to Fleishman,* 8, and Graves, *A View*, 249.

16. Deaton oral history, Hunter.

17. Gott, *A Tribute to Fleishman,* 9.

18. Deaton oral history, Tanner.

Chapter 6: Class 43-2

1. Margaret Ann Hamilton Tunner, WASP Class 43-2, oral history conducted by Dr. Deborah G. Douglas, May 8, 1998, held in The H.O. Malone collection held in the WASP Archives at TWU.

2. Graves, *A View from the Doghouse*, 134.

3. Elizabeth Whitlow Rummerfield Smith, WASP Class 43-2, an oral history, interviewed by Dawn Letson, August 4, 1999, TWU.

4. Mary Catharine "Jary" Johnson McKay, WASP Class of 43-W-2, an oral history, interviewed by Dawn Letson, October 14, 1994, TWU.

5. Catherine "Cappy" Vail Bridge, WASP Class 43-W-2, an oral history, interviewed by Dawn Letson, October 27, 2003, TWU.

6. Marie Muccie Genaro, WASP Class 43-2, an oral history, interviewed by Cindy Weigand, July 12, 2002. TWU.

7. Ruth Dailey Helm, WASP Class 43-2, an oral history, interviewed by Tracey MacGowan, October 3, 2002, TWU.

8. Martha Wagenseil (Davis Rupley), WASP Class 43-2. Transcript of a taped oral history recorded by Marty, herself, probably in the 1980s and

given to WASP author Byrd Granger. A transcript of the tape is part of the Granger collection, TWU WASP Archives. Copy in the author's hands.

9. Iris Cummings Critchell, WASP Class 43-2. Personal comments by Critchell to the author based on her memoir-in-progress, working title "76 Years With the Spirit of Flight."

Chapter 7: Class 43-3

1. Lois Hollingsworth Ziler, WASP Class 43-3, an oral history conducted by Dawn Letson, May 21, 1994, TWU.

2. Lois Brooks Hailey, WASP Class 43-3, an oral history conducted by Dawn Letson, May 21, 1994, TWU.

3. Form 64 Army Air Forces physical exam.

4. Margaret Sanford Oldenburg (43-4) was the first training casualty of the WFTD (later WASP) training program. She was killed, along with her instructor, Norris G. Morgan, when their PT-19 crashed at Minnetex, seven miles south of Houston on March 7, 1942. They were on a routine training flight. Two sources: *In Memoriam: Thirty-eight American Women Pilots* (Denton: Texas Woman's University Press, 1996), researched and written by Dawn Seymour, 43-5; Clarice I. Bergemann, 44-2; Jeanette J. Jenkins, 44-1; and Mary Ellen Keil, 44-2; see page 27. Also, Celeste Graves, *A View from the Doghouse*, vii.

5. Lillian Connor Roberts-Risdon, WASP Class 43-3, an oral history conducted by Dawn Letson, October 7–8, 1996, TWU.

6. http://en.wikipedia.org/wiki/Army_Air_Forces_Training_Command; accessed July 10, 2011. Army Air Corps Flying Training Command was constituted and established on January 23, 1942. Its mission was to train pilots, flying specialists, and combat crews. Redesignated as Army Air Forces Flying Training Command on or about March 15, 1942, after the Army Air Forces became a subordinate but

autonomous arm of the United States Army. In July 1943 this command merged with the AAF Technical Training Command to form the Army Air Forces Training Command.

7. The United States Department of War, at the beginning of World War II, was headquartered in the Munitions Building, a temporary structure erected during World War I along Constitution Avenue on the National Mall. http://en.wikipedia.org/wiki/The_Pentagon; accessed 7-10-2011.

8. http://en.wikipedia.org/wiki/Army_Air_Forces_Training_Command;

9. Deaton oral history, Hunter.

10. Deaton oral history, Hunter.

11. Lois Hollingsworth Ziler interview.

Chapter 8: Classes 43-4, 43-5, 43-6

1. Jean Landis, 43-4, oral history, May 13 and September 16, 2012. Interviewer Sarah Byrn Rickman for WASP Archives, TWU.

2. Ibid.

3. Dorothy Kocher Olsen, 43-4, oral history, May 2005. Interviewer Sarah Byrn Rickman for the WASP Archives, TWU.

4. Virginia "Virgie" Lee Jowell Hagerstrom, Class 43-4, oral history, October 13, 1994, interviewer Dawn Letson.

5. Alice Jean "A.J." May Starr, Class 43-4, oral history, February 27, 2011. Interviewer Sarah Byrn Rickman, for the WASP Archives at TWU.

6. Betty Stagg Turner, *Out of the Blue and Into History* (Arlington Heights, IL: Aviatrix Publishing Inc., 2001), 123-124.

7. Starr, oral history, TWU.

8. *On Silver Wings,* Vi Cowden DVD; Turner, *Out of the Blue,* 125.

9. Gayle Bevis Ewing Reed, 43-5, an oral history for TWU, interviewed by Dawn Letson October 2, 2004.

10. Sylvia Dahmes Clayton, 43-5, an oral history for TWU, interviewed by Dawn Letson, October 5, 2002.

11. Adeline Wolak Ellison, 43-6, an oral history for TWU, interview by Lindsey Wright, September 30, 2004.

12. Turner, *Out of the Blue,* 209.

13. Anna Flynn Monkiewicz, 43-6, an oral history for TWU, interviewed by Dawn Letson, October 2006.

14. Turner, *Out of the Blue,* 184.

15. Turner, *Out of the Blue,* 182.

16. Deaton oral history, Hunter, TWU.

Part III: Three More Ferrying Squadrons Formed *January–June 1943*

1. Memorandum from Maj. Hamilton Heard to Col Robert Love, March 25, 1943.

Chapter 9: 5th Ferrying Group, Love Field, Dallas, Texas

1. George W. Cearley, Jr., *Dallas Love Field* (A Pictorial History of Airline Service, self-published, 1989), 5. Dallas Love Field Home Page: Love Notes, Chronology of Events and Love Field Facts: www.dallaslovefield.com/lovenotes/lovechrono.html

2. Frontiers of Flight Museum, Love Field website: http://www.flightmuseum.com/about.htm

3. Marion Stegeman Hodgson, *Winning My Wings* (Albany, TX: Bright Sky Press, 2005), 147.

4. Rickman, *The Originals*, 122, 379–80.

5. Sarah Byrn Rickman, "Born to Fly," *Kansas Heritage*, a magazine of the Kansas State Historical Society, Winter 2006. This is an article about WAFS Betsy Ferguson.

6. Information collected from the Dorothy Scott Collection—her letters home—held at TWU, the Woman's Collection, WASP Archives.

7. Bohn, "Catch a Shooting Star," Chapter 9, 11–12.

8. Bohn, "Catch," Chapter 10, 13. Bohn writes a description of Nancy's flight based on a letter she received from Nancy.

9. Nancy Love's logbooks—NHL. Nancy flew a total of three hours and fifteen minutes in the P-51 over a three-day period, February 27, March 1, and 2. Copies in author's collection.

10. Delphine's personal papers. Author has copies.

11. Helen Richards's personal papers, including her documented flight hours as a WAFS. Copies are held at Texas Woman's University in the WASP Archives and by the author.

12. Celeste Graves, *A View from the Doghouse* (Bloomington, IN: AuthorHouse, 2004), 19, 23, 28, 33, 49.

13. Byrd Howell Granger, *On Final Approach*, 126.

14. Marx, 394.

15. Granger, *On Final*, 126.

16. Granger, *On Final*, pp. A-42/B, A-43/B, A-44/B. These three pages list what appears to be all of the women pilots who served in Dallas for any length of time—be it only a few days.

Chapter 10: 2nd Ferrying Group, New Castle AAB, Wilmington, Delaware

1. Sources for Betty Huyler Gillies: *Catch a Shooting Star*, Chapter IV, 3–4, by Delphine Bohn; Betty Gillies private papers, International Women's Air and Space Museum, Cleveland; letters from Betty Gillies to Delphine Bohn in answer to a questionnaire, The Woman's Collection, TWU; and personal interview with Betty Gillies by Dawn Letson, TWU, September 1996. All these sources were used by this author, Sarah Byrn Rickman, in her book *The Originals: The Women's Auxiliary Ferrying Squadron of World War II. The Ninety-Nines, Yesterday—Today—Tomorrow* (Paducah, KY: Turner Publishing Company, 1996), 11-12.

2. Taped interview with Eduardo Escallon, Helen's nephew, February 2002, and her World War II papers, saved by her sister, Roberta, Eduardo's mother. The tape is held in the WASP Archives at TWU. Both the author and TWU hold copies of the papers.

3. Sam Parker, an article on Sis Bernheim in the *Journal, American Aviation Historical Society,* summer 1995. Copy held by WASP Archives, TWU.

4. Rickman, *Nancy Batson Crews,* 24–27.

5. Rickman, *The Originals*, 176, 354. Original source, an audiotaped interview with W. Gerould Clark III, held in the WASP Archives, TWU, and a 2001 phone interview with Clark and his brother William Goodman Clark.

6. Rickman, *The Originals*, 361–362. The author's audiotaped interview with Dorothy's sister, Honey Fulton Parker, is held by the WASP Archives, TWU.

7. Rickman, *The Originals*, 176, 371–372. Gertrude's oral history, as well a videotaped interview done by the author in 1999, are held by the WASP Archives, TWU.

8. Rickman, *The Originals,* 35, 37, 46, 366. Original source, the author's several one-on-one unrecorded interviews in March 2000, at Teresa's home in Lake Worth, Florida.

9. Jan Churchill, *On Wings to War: Teresa James, Aviator* (Manhattan, KS: Sunflower University Press, 1992), 32.

10. Rickman, *The Originals,* 375–376.

11. Rickman, *The Originals,* 128–129.

12. Information and quotes taken from Betty Gillies's letter to Delphine Bohn dated January 8, 1984, and from Betty Gillies, oral history, TWU.

13. The 2000-horsepower Republic P-47 Thunderbolt was very effective as a short-to-medium range escort fighter in high-altitude air-to-air combat. http://rwebs.net/avhistory/history/p-47.htm; http://en.wikipedia.org/wiki/Republic_P-47_Thunderbolt. Accessed Oct. 18, 2013.

14. Author's personal collection includes copies of orders dated August 6 and 23, and October 7, 1943, listing Teresa James as the pilot charged with delivering a specified P-47D from Republic Aircraft Corp. to a stated location. These would be among her earliest P-47 deliveries. She notes on the August 6 orders, "You know I should frame this—my first P-47 trip." Betty Gillies is listed on the August 6 orders also delivering a P-47D.

15. Graves, *The View from the Dog House,* 6.

16. Graves, *The View,* 65.

17. Graves, *The View,* 25.

18. Graves, *The View,* 44.

19. Graves, *The View,* 63.

20. Marie Muccie Genaro, oral history.

CHAPTER 11: 3RD FERRYING GROUP, ROMULUS, MICHIGAN

1. Daniel W. Mason, *Detroit Metro Airport, The Images of Aviation* (Charleston, SC: Arcadia Publishing, 2011), 7.

2. Willow Run# from Wikipedia, the free encyclopedia.

3. Mason, *Detroit Metro Airport*, 27.

4. Sarah Byrn Rickman, "Women with Wings: The WASP in Michigan," *Michigan History* magazine, published in July/August 2010.

5. Turner, *Out of the Blue*, 15.

6. Rickman, *The Originals*, 124.

7. *The Originals*, 378–379.

8. *The Originals*, 124–125.

9. *The Originals*, 352–353.

10. Iris Critchell, phone conversation with author, August 10, 2013.

11. *The Originals*, 125–126, 385–386.

12. *The Originals*, 127.

13. LaFarge, "Women Pilots," 82–83.

14. Chief of Staff ATC Gen. C.R. Smith to Commanding Office, Ferrying Division, dated April 17, 1943, Subject: WAFS; LaFarge, "Women Pilots of the ATC," Appendix II, 66, in TWU, WASP Archives.

15. LaFarge, "Women Pilots," 82–83.

16. Tunner letter of April 26, 1943, NHL collection. Also LaFarge, "Women Pilots of the ATC," Appendix II, 66, in TWU, WASP Archives.

17. Marx, "Women Pilots Ferrying Division," 115–116.

18. Marx, 111.

19. Graves, *A View from the Doghouse*, 8, 13, 35, 47, 48.

20. Scharr, *Sisters in the Sky*, vol. 1, 446.

21. Scharr, 454.

22. Scharr, 462.

23. Scharr, 477–482.

24. Scharr, 505, 510.

25. Granger, *On Final Approach*, 134.

Chapter 12: 6th Ferrying Group, Long Beach, California

1. History of the Sixth Ferrying Group.

2. Sarah Byrn Rickman, "Barbara Erickson: From 'Rosie the Riveter' to B-17 Pilot," *Air Power History*, Summer 2005.

3. Rickman, *The Originals*, 387–88.

4. Rickman, *The Originals*, 349.

5. Rob Simbeck, *Daughter of the Air: The Brief Soaring Life of Cornelia Fort* (New York: Atlantic Monthly Press, 1990), 227–28.

6. Simbeck, 230.

7. Simbeck, 232.

8. Sarah Byrn Rickman, *Nancy Love*, 118.

9. Simbeck, 234–35.

10. LaFarge, "Women Pilots in the ATC," 40.

11. Graves, *A View from the Doghouse*, 30, 37, 52.

12. Critchell, conversations with author and excerpts from her memoir.

Chapter 13: Welcome to the Ferry Command

1. Marx, 195–196, paraphrasing letter from General George to Colonel Tunner, dated May 6, 1943, subject "Women's Training Program."

2. Marx, 197, paraphrasing letter from Colonel Baker to Colonel Tunner dated May 12, 1943.

3. Jane Straughan, oral history, TWU.

4. Marx, 199-200. Letter from Nancy Love to General Yount, June 26, 1943.

5. The Stearman became the primary trainer in time for Class 44-5 to take all of its primary training and that class entered training December 7, 1943. However, Class 44-4—that entered training November 1, 1943—prior to graduation May 23, 1944, was given additional hours in the PT-17 to give them experience in the narrow-geared trainer. Information from interviews with Mildred Jane Baessler Doyle 44-4, in person, September 22, 2012, and with Florence "Shutsy" Reynolds, 44-5, via email, October 26, 2012.

6. Marx, 200. Letter, Love to Yount.

7. "Nancy Love Visits Avenger Field," *The Avenger* (base newspaper), June 1943. Copy from Margaret H. (Reeves) Riviere (WASP Class 43-4), private collection. Interview conducted March 31, 2006, in Barnesville, Georgia, on file in WASP Archives, TWU.

Part IV: Missions Altered *July-November 1943*

1. *World War II Day by Day,* 424.

2. *World War II Day by Day,* 440, 442-43.

3. *World War II Day by Day, 463.*

4. *World War II Day by Day, 462.*

Chapter 14: Cochran's Power Play

1. Eisenhower Library, Memo from Cochran to Arnold, May 3, 1943.

2. AAF Memorandum No. 20-4. LaFarge, "Women Pilots of the ATC," Appendix II, 69-70.

3. Letter from Major Gordon A. Rust, Ferrying Division/ATC, to Hazel Taylor, WAFS Publicity, Bureau of Public Relations, War Department, Washington D.C., dated June 15, 1943, held in Archambault / Jernigan collection BHG F3, #2, TWU.

4. Memorandum from General Tunner to Commanding Officers all Ferrying Groups Subject: Duties of Executive for WAFS. LaFarge, "Women Pilots of the ATC" Appendix II, 71-72.

5. Tunner memorandum, LaFarge Appendix.

6. Marx, 133; #55, "Women Pilots AAF," 44.

7. Source: memo from FD to ATC, July 26, 1943: "Directives Regarding Women Pilots," found in Tunner's files, Ferrying Division. LaFarge, 98-99, including fns 1–4.

8. #55, "Women Pilots AAF," 48, fn 30.

9. Complete text of a report by Jacqueline Cochran, Director of Women Pilots, AAF, to the Commanding General of the AAF, General H. H. Arnold, together with a summery of the report, both for release to the press Tuesday A.M., August 8, 1944. Copy held by TWU, WASP Archives in the Jernigan/Archambault collection and elsewhere. In the document under "E—How Trained WASPs are Serving," first subsection #1. Under #4, Cochran also states re: target towing work done by the WASP: "Target towing is work that most male pilots dislike and, consequently, they take little interest in doing a good job. Further, such flying affords little opportunity for experience [that] would be of value as pre-combat training."

10. Keil, *Those Wonderful Women*, 352.

11. Douglas, *American Women and Flight,* 87. Douglas cites interviews with WAFS Betty Gillies and Delphine Bohn and Love's daughter, Margaret Love, as well as Mardo Crane's "The Women With Silver Wings," 10.

12. Rich, *Pilot in the Fastest Lane,* 130.

13. Marx, "Women Pilots," 151-152. Footnote #15 reads: "Letter from Hq, ATC, to CG, FD, July 17, 1943, Subject: Assignment of 25 Women Pilots." CG is General Tunner, FD is Ferrying Division.

14. Lois Brooks Hailey, oral history.

15. Granger, *On Final Approach,* 139–140.

16. Rickman, *The Originals,* 182–183.

17. Scharr, *Sisters in the Sky,* vol. 1, 519.

18. Granger, A-57 / B.

19. Scharr, 522.

20. Scharr, 524.

21. Marx, "Women Pilots," 133.

22. Douglas, *American Women in Flight,* 89.

23. From Betty's speech given to The Ninety-Nines Inc., International Convention Banquet, Baltimore MD, June 27, 1985, and to the Southern California WASP, Laguna Hills, CA, June 14, 1987. From Betty Gillies Collection, IWASM.

Chapter 15: Nancy and Betty Fly the Fortress

1. Tunner, 37–38.

2. Betty Gillies, oral history, WASP Archives, TWU.

3. Tunner, 38.

4. Philip Makanna and Jeffrey Ethell, *Ghosts: Vintage Aircraft of World War II* (Charlottesville, VA: Thomasson-Grant, 1989), 111.

5. Letter from Colonel George D. Campbell, Director of Operations and Training, Ferrying Division, to the Commanding General re: qualification of Nancy and Betty as B-17 crew, August 16, 1943, Nancy Harkness Love personal collection, in the hands of her daughter Margaret Love. The author also has a copy.

6. Telex dated September 4, 1943, from Gen. C.R. Smith to General Paul Burrows and Major Roy Atwood, copy held at TWU.

7. Cable from General Arnold in London to Presque Isle Air Base, September 5, 1943, TWU.

8. Rickman, *Nancy Love*, 144–145.

9. Gillies, oral history.

10. Elinor Fairchild Stebbins, 44-6, oral history, May 18, 2006, interviewer, Sarah Byrn Rickman, WASP Archives, TWU. Elinor Stebbins and Dora Dougherty Strother McKeown, 43-3, were assigned as copilots on B-17s headed for Puerto Rico in December 1944, on orders from none other than Paul Tibbetts, the pilot who later flew the *Enola Gay* and dropped the atomic bomb on Hiroshima. The AAF was moving the "prepare for overseas movement" (POM) base from Nebraska to the more welcoming winter climate in Puerto Rico. From Stebbins's oral history, WASP Archives, TWU.

11. Barbara Erickson London, as told to Sarah Byrn Rickman, "The Lady Flew the B-17," *Aloft, The Museum of Flight Magazine,* September/October 2005.

Chapter 16: New Mission for the Ferrying Division

1. Tunner, *Over the Hump,* 27–29.

2. Iris Critchell, phone conversations with the author, spring 2015.

3. Tunner, 27–28.

4. Marx, 118, staff meeting notes April 28, 1943.

5. Marx, 119, staff meeting notes June 30, 1943.

6. Marx, 207.

7. Two documents contain in-depth discussions from which information and direct quotes in the next several paragraphs were taken. The author refers interested readers to: LaFarge, 126-128 and #55, "Women Pilots AAF," 60-62.

8. LaFarge, 127,

9. Tunner, 27–28.

10. Letter from Tunner to Commanding General ATC, attention Col. Robert M. Love, following an October 6 conference, Subject: WASP. Copy held in BHG F3 #1, Archambault/ Jernigan, TWU WASP Archives.

11. #55, "Women Pilots AAF," 61; LaFarge, 128.

12. #55, "Women Pilots AAF," 59; and LaFarge, "Women Pilots ATC," 126: reference Qualifications of Women Pilots Assigned to the FD/ATC, September 14, 1943.

13. #55, "Women Pilots AAF," 62.

14. LaFarge, 126.

15. #55, "Women Pilots AAF," 62-63. Marx, 207, Letter FD to ATC, Sept. 21.

16. Letter from Tunner to ATC following an October 6 conference, Subject: WASP.

17. #55, "Women Pilots AAF," 59.

18. Natalie Jeanne Stewart-Smith: "An Appraisal of AAF Training of Individual Specialists, 1941–45," page 2 of the introduction, as quoted in: "The Women Airforce Service Pilots (WASP) of World War II: Perspectives

on the Work of America's First Military Women Aviators," an abstract of Master's thesis done by Stewart-Smith, Washington State University, 1981, page 35. The document is held by TWU, the WASP Archives.

19. The percentages are the author's, checked by her mathematician husband.

20. *World War II Day by Day*, 467.

Chapter 17: The Air Inspector

1. #55, "Women Pilots AAF," 59.

2. "Cubstuff": slang, all-encompassing terminology for single-engine trainer aircraft like the J-3 Cub, manufactured by Piper Aircraft, and others built by Taylorcraft, Aeronca, Ryan, and Stinson, etc.

3. #55, "Women Pilots AAF," 59.

4. #55, "Women Pilots AAF," 60.

5. #55, "Women Pilots AAF," 59.

6. Telegram to Jacqueline Cochran from Class W-4, September 22, 1943, BHG F3 #1, (310), Archambault/Jernigan collection, Bolling AFB, Washington DC. Also Granger, 182.

7. Letter from Tunner to Commanding General ATC, September 14, 1943, BHG F3 #2, (1269-70), Archambault/Jernigan collection, Bolling AFB, Washington DC.

8. Memo dated September 18, 1943, from C.R. Smith, deputy commander, ATC, to Asst. Chief Air Staff, Training, Subject: Lack of Flight Qualifications, WASP, BHG F3 #2, (1268).

9. Letter from Jacqueline Cochran to Air Inspector, dated September 29, 1943, Subject: Lack of Flight Qualifications, WASP, BHG F3 #2, (1264-65).

10. Letter from Gen. Robert W. Harper to Air Inspector, dated October 1, 1943, Subject: Lack of Flight Qualifications, WASP, BHG F3 #2, (1263).

11. WASP Elimination Utilization of Women Pilot Graduates Resume of Report: Dudley N. Outcalt, Lt. Colonel, A.C., Investigating Officer, BHG F3 #2 (1271-1295). This is inclusive of several paragraphs.

Chapter 18: ATC Holds Firm in Tussle for Command

1. #55, "Women Pilots AAF," 27-29.

2. #55, "Women Pilots AAF," 31.

3. LaFarge, 110; Marx, 164-165; Verges, 150; Granger, 201-202.

4. LaFarge, 111–112, see footnote 30, dated November 3, 1943; Verges, 150.

5. LaFarge, 114; Marx, 168.

6. LaFarge, 112–113, footnotes 31 and 32.

7. LaFarge, 114, also reference footnote 33, which refers back to footnote 31 on 112. Also Marx, 168.

8. LaFarge, Notes on interview with Mrs. Nancy H. Love, 3 October 1945, TWU.

9. Marx, 166, fn 27, letter from George dated November 3, Subject: Regulations and Organization of WASP.

10. Verges, 151.

11. LaFarge: The Ferrying Division's proposed regulation—counter to Cochran's proposal—can be found, in its entirety and dated November 5, 1943, in Appendix II of the LaFarge official history, 78-81. Copy available at TWU, part of the Archambault / Jernigan research, Bolling AFB, Washington D.C.

12. LaFarge. The final AAF Regulation No. 40-8, dated December 21, 1943, can be found in the same volume, 87–91, signed by Maj. Gen. Barney M. Giles, Chief of Air Staff.

13. Marx, 171.

14. LaFarge, 115.

15. Verges, *On Silver Wings,* 134.

16. LaFarge, 115.

17. LaFarge, *The Eagle in the Egg,* 133–134.

18. LaFarge, *The Eagle in the Egg,* 134.

Chapter 19: Ten WASP Ferry Pilots Tell All!

1.Iris Cummings Critchell, memoir-in-progress, working title "76 Years With the Spirit of Flight."

2. Jane Straughan, oral history, TWU.

3. Ruth Dailey Helm, oral history, TWU.

4. Mary C. "Jary" Johnson McKay, oral history, TWU.

5. Marion Schorr Betzler Brown, oral history, TWU.

6. Pilotage is the use of fixed visual references on the ground or sea by means of sight or radar to guide oneself to a destination, sometimes with the help of a map or nautical chart. http://en.wikipedia.org/wiki/Pilotage, accessed August 10, 2012.

7. Dorothy Kocher Olsen oral history, TWU. Addendum, Iris Critchell even checked her logbook to fill in some of the extra details of that A-24 flight. Email to the author, September 15, 2013.

8. Gayle Bevis Ewing Reed oral history, TWU.

9. Sylvia Dahmes Clayton oral history, TWU.

10. Inez Woodward Woods, WASP 43-4, two oral histories, WASP Archives, TWU. One done by Sarah Rickman, September 20, 2005, and

one done by Ruth Florey, for the Confederate (Commemorative) Air Force Library, Midland-Odessa, Texas. No date given.

11. The PQ-14 was a radio-controlled target drone for training anti-aircraft artillery gunners made by Culver Aircraft Company. It had retractable landing gear and fuselage, wings and tail components made of wood with stressed plywood skin. http://en.wikipedia.org/wiki/Culver_PQ-14_Cadet – accessed 8-8-2012

12. Placard: a small card or metal plaque, in this case delineating a prohibited maneuver in that specific aircraft. Aviation's definition found in *FAR/AIM, Federal Aviation Regulations and Aeronautical Information Manual* (U.S. Department of Transportation, Newcastle, WA: Aviation Supplies & Academics Inc. or ASA, 2011), 162.

13. Maneuvers: A military exercise (also called war games) is the employment of military resources in training for military operations, either exploring the effects of warfare or testing strategies without actual combat. http://en.wikipedia.org/wiki/Military_operation—accessed 9-6-2012.

14. Aircraft dope is a plasticized lacquer that is applied to fabric-covered aircraft. It tautens and stiffens fabric stretched over airframes and renders them airtight and weatherproof. http://en.wikipedia.org/wiki/Aircraft_dope, accessed Aug. 8, 2012

15. A.J. May Starr, oral history, TWU. Also an email from A.J. dated September 13, 2012, containing additional detail.

16. Granger, *On Final*, 192.

Part VI

1. Dorothy Scott (WAFS) Collection, TWU, Dorothy Scott's letters home, letter dated December 1, 1943, TWU.

2. *World War II Day by Day*, 464.

3. *World War II Day by Day,* 466.

4. *World War II Day by Day,* 468.

5. Ploesti was a vast complex of oil refinery facilities located some thirty miles north of Bucharest, Romania. It supplied an estimated 60 percent of the refined oil necessary to keep the German war machine running. http://www.eyewitnesstohistory.com/ploesti.htm, accessed Feb. 5, 2015.

6. Barrett Tillman, *Forgotten Fifteenth* (Washington D.C.: Regnery History/ Regnery Publishing, 2014), 56–57.

Chapter 20: Pursuit School

1. Iris Cummings Critchell, 43-2. Material from her memoir-in-progress, working title "76 Years With the Spirit of Flight."

2. SNAFU: situation normal, all fouled up; FUBAR, fouled up beyond all recognition; TARFU, things are really fouled up. If you wish, you can insert another "F" word.

3. Anna Flynn Monkiewicz (WASP 43-6) *My Piece of the Sky* (Amazon Digital Services, Inc., April 4, 2013.) Through Kindle, I have this book on my iPad. The text quoted is excerpted from the second and third paragraphs of Chapter Twelve, "Back to Romulus."

4. BC-1—"BC" stood for basic combat and it was the first model of what became the advanced single-engine trainer in WWII, the AT-6 Texan. The USAAF received 180 BC-1s. http://en.wikipedia.org/wiki/North_American_T-6_Texan accessed 9-16-2013. From entries in logbooks of WASP who attended pursuit school, it appears that the BC-1 was used for pursuit training in Palm Springs and the AT-6 was used for pursuit training in Brownsville.

5. The radial engine is a reciprocating-type internal combustion engine configuration in which the cylinders point outward from a central crankshaft like the spokes of a wheel. Radial engines are regarded as

being air-cooled almost by definition. Liquid-cooled engines often weigh more and their cooling systems are both more complex and are generally more vulnerable to battle damage. Minor shrapnel damage easily results in a loss of coolant and consequent engine seizure, while an air-cooled radial would be unaffected. http://en.wikipedia.org/wiki/Radial_engine. Accessed Nov. 9, 2012.

6. Dorothy Scott letters home, TWU.

7. Catherine Vail Bridge, oral history.

8. Elizabeth Whitlow Smith, oral history.

9. Gott, *Women in Pursuit,* 86

10. Gott, *Women in Pursuit.* The listings and photos of these thirty-two WASP Pursuit School graduates are found on 75-81.

11. Jean Landis, 43-4, oral history.

Chapter 21: Militarization

1. Marx, "Women Pilots Ferrying Division," 295–296.

2. LaFarge, "Women Pilots ATC," 161.

3. Col. Oveta Culp Hobby, Director of the Women's Army Auxiliary Corps (WAAC) and later the Women's Army Corps (WAC).

4. Jacqueline Cochran Oral History, United States Air Force Academy, # 44, March 11-12, 1976, 39-41.

5. #55, "Women Pilots AAF," 85.

6. #55, "Women Pilots AAF," 85–86.

7. LaFarge, "Women Pilots ATC," Appendix II Documents, full text, 82–83.

8. LaFarge, "Women Pilots ATC," 165; Marx, "Women Pilots Ferrying Division," 290–291.

9. "In wing and higher level organizations, A-1 is the staff officer concerned with policies and plans relating to personnel." In November 1943, Maj. Gen. J. M. Bevans, Assistant Chief, Air Staff, personnel, held that post. Thank you to members of the staff at the National Museum of the United States Air Force for the explanation.

10. Archambeault/Jernigan collection, TWU, Vol. 1 of 5 (BHG F3 #1); Comments of Director of Women Pilots with respect to the historical report of Ferrying Division Air Transport Command concerning WASP in said division, 59, quoted in Captain Marx's report.

11. Ibid., 59–60.

12. #55, "Women Pilots AAF," 91–93; Marx, "Women Pilots Ferrying Division," 298–299.

13. Fahey, *U.S. Army Aircraft 1908–1946*, 21. A-30 twin-engine attack aircraft built by Martin in Baltimore primarily for the British. Orders for 900 of these aircraft subsequently were cancelled.

14. Betty Gillies's five-part memo to Delphine Bohn dated from November 29, 1983, to February 8, 1984. A copy of the memo is held by the WASP Archives, TWU. Betty answered a series of questions from Delphine. She recalled events using notes in the diary she kept throughout her WAFS/WASP service. Specific information given about the dates of January 5–10, 1944, is used in this chapter.

15. #55, "Women Pilots AAF," 88.

16. #55, "Women Pilots AAF," 88–89.

17. Betty Gillies's memo to Delphine Bohn.

18. #55, "Women Pilots AAF," 89–90.

19. Esther Manning Rathsfelder diary from 1944, in the hands of her daughter, Julie Shively. Shared with the author.

20. Bohn, "Catch," Chapter 15, 6.

21. Bohn, "Catch," Chapter 15, 5–7.

22. Barbara London, phone interview, July 7, 2004.

23. *World War II Day by Day,* 492.

24. *World War II Day by Day,* 494.

25. #55, "Women Pilots AAF," 91.

26. Verges, *On Silver Wings,* 188. Also Barbara London.

27. #55, "Women Pilots AAF," 94.

28. Marx, 312.

29. Keil, *Those Wonderful Women*, 302–3.

30. Molly Merryman, *Clipped Wings* (New York: New York University Press, 1998), Chapter 4, "From Praise to Rancor." Also Rickman, *Nancy Love,* 201-4.

31. Marx, 313.

32. Marx 314, fn 27.

33. Marx, 315, fn 28.

34. Marx, 313–317.

35. Marx, 317.

36. Marx, 322.

Chapter 22: Officer Training School

1. Marx, "Women Pilots in the Ferrying Division," 292. Captain Marx is using as his source the Ferrying Division Staff Meeting Notes dated March 13, 1944, and the Weekly Summary dated March 29, 1944.

2. Marx, 292–93.

3. Marx, 317.

4. Granger, *On Final Approach,* 384-86, 389.

5. Marx, 318–319 and including fn. 29.

6. Report of the Air Surgeon's Office on WASP Personnel, 1945, 27. From the files of the Air Force Museum, Wright Patterson AFB, Ohio, Research Dept. File L2, "Wom Air For Serv Pilots," loaned to the author by Dr. Jacque Boyd, Aero Infosearch.

7. Letter from WASP Ruth Roberts, Nancy Harkness Love collection, TWU.

8. Diane Bartels, *Sharpie: The Life Story of Evelyn Sharp, Nebraska's Aviatrix* (Lincoln, NB: Dageforde Publishing, 1996).

9. Bartels, 223.

10. Bartels, 236–43, 8–10, 247.

11. http://en.wikipedia.org/wiki/P-38_Lightning#Specifications_.28P-38L.29. Accessed Jan. 23, 2012. "It was the first American fighter to make extensive use of stainless steel and smooth, flush-riveted butt-jointed aluminum skin panels. [20] It was also the first fighter to fly faster than 400 mph (640 km/h)."

12. Marx, 307.

13. Granger, 320.

14. Barrett Tillman, *Forgotten Fifteenth,* 56–57.

15. Gott, *Women in Pursuit,* 81.

16. Gott, 81–84.

Chapter 23: The First Transfers

1. Marx, 306–7.

2. #55, "Women Pilots AAF," 62–63.

3. Marx, 242. Quoting General Tunner referencing General George's letter dated October 16, 1943.

4. Marx, 236-237. Quoting memo to General Tunner from Mrs. Love, December 14, 1943.

5. Marx, 238. Quoting Letter, January 8, 1944—WASP Training.

6. Marx, 239–241. Quoting Letter, January 10, 1944—WASP.

7. IRIS # 145345, WASP Activities in the Air Training Command Volume 3. Air Force Historical Research Agency, Department of the Air Force, Maxwell AFB, Alabama. Letter to Miss Edwards, from Eldred F. Graham, Headquarters, Ferrying Division, Air Transport Command, Cincinnati 2, Ohio, dated April 26, 1944. The letter is a verification of payroll changes.

8. http://www.airspacemag.com/military-aviation/kansas-city-b-25-factory-180951624/?no-ist, accessed Nov. 8, 2015.

9. Marie Muccie Genaro, oral history, TWU.

10. Granger, A-30 / B.

11. Marx, 222, Second Indorsement [*sic*] to May 25 letter to Arnold, Training Command, from HDQ AAF, June 17.

12. Nancy Harkness Love deposition, found in the Pat Pateman Collection, TWU. Information is also found in the author's biography, *Nancy Love and the WASP Ferry Pilots of WWII*, 181–82.

13. Esther Manning 1944 diary.

Chapter 24: 21st Ferrying Group, Palm Springs

1. Granger, *On Final Approach*, A-53/B-A-54/B. The WAFS were: Evelyn Sharp, Helen Richards, Esther Nelson, Dorothy Fulton, Bernice Batten, and Katherine Rawls Thompson. The WFTDs were Zelda Lamer, Ellen Gery, Claire Callaghan, Ruby Mullens, Mary Lou Colbert, Ruth Franckling, Helen Porter, Barbara Russell, Vega Johnson, and Martha Wagenseil.

2. Gott, *Women in Pursuit*, 76-81.

3. Granger, 300–301.

4. Granger, 317–19.

5. Author interview with Neale, April 7, 2007.

6. Martha Wagenseil oral history transcript, TWU, used throughout the chapter.

7. Martha Wagenseil, oral history.

8. Cappy Vail Bridge, oral history.

9. Vivian Cadman Eddy, oral history.

10. Commendation for WASP Squadron, 21st FG, ATC 10 October 1944, signed Byrd H. Granger, Squadron Leader, copy in Granger collection, TWU.

Chapter 25: D-Day, the Sixth of June

1. Told to author by Margaret Ann Hamilton Tunner, while visiting her at Hockley Farm in Virginia, June 2003; additional details from Granger, *On Final*, 372.

2. Rickman, *The Originals*, 258–260, based on the author's interviews with Nancy Batson in 2000.

3. Mary (Jary) Johnson McKay, WASP 43-2, oral history. She relates the details of Dottie Nichols's death.

4. Granger, *On Final Approach*, 326–327.

5. McKay, oral history.

6. Granger, *On Final Approach*, 327. Nancy Love's logbooks confirm that she was at Long Beach in the June 20-21 time period. Author's collection.

7. Gott, *Women in Pursuit*, 82, 168.

8. Col. Ponton D'Arce became the Commanding Officer of the 3rd Ferrying Group, Romulus, Michigan, in May 1943.

9. McKay, oral history.

10. Marx; a full explanation of the Ramspeck Report is found on pages 320–331.

11. Granger, Rep. Joseph O'Hara, Minnesota, 343.

12. Marx, 334, material taken from the "Congressional Record" June 21, 1944, Vol. 90, No. 116, 6487.

13. Marx, 337.

14. Marx, "Women Pilots Ferrying Division," 322–323, under fn 31, 320. House Report No, 1600, Concerning inquiries made of certain proposals for the expansion and change in civil service status of the WASP," June 5, 1944.

15. Marx, "Women Pilots Ferrying Division," 328-329, under fn 31, 320.

16. Marx, Ferrying Division staff notes, June 30, 1944, 337.

Chapter 26: A Typical Coast-to-Coast P-51 Delivery

1. Mary Catherine "Jary" Johnson, WASP 43-2, oral history, 21.

2. Iris Cummings Critchell, WASP 43-2; material from her memoir-in-progress, working title "76 Years With the Spirit of Flight."

3. San Gorgonio Pass: The San Gorgonio Mountain (11,503 feet) is at the pass's northern end, and San Jacinto Peak (10,834 feet) is at the southern end. The peak is only six miles south of Interstate 10 that, today, cuts through the pass. http://www.absoluteastronomy.com/topics/San_Gorgonio_Pass. http://www.absoluteastronomy.com/topics/San_Gorgonio_Mountain.

http://www.absoluteastronomy.com/topics/San_Jacinto_Peak. Accessed Dec. 30, 2011.

4. Iris Cummings Critchell, material from her memoir-in-progress, working title "76 Years With the Spirit of Flight."

5. *Wings of Silver: The Vi Cowden Story* (WASP class 43-4) a 33-minute DVD. Copyright 2010, The Monkey House, California.

6. Email from Iris Critchell to the author dated May 24, 2011, in response to a specific question about the WASP ferry pilots carrying sidearms.

7. Author interviews with Nancy Batson Crews, recorded in June, August, November, and December 2000. In author's possession.

Part VII

1. *World War II Day by Day*, 544.

2. *World War II Day by Day*, 546.

3. *World War II Day by Day*, 549.

4. *World War II Day by Day*, 549.

Chapter 27: *The Great Transfer*

1. Hodges, *Winning My Wings*, 147.

2. Gott, *Women in Pursuit*, 198.

3. "Nancy H. Love Visits Field, Sees Largest Class of Pilots Graduate," Ferrying Division/Air Transport Command base newspaper, June 10, 1944, in NHL collection.

4. Transcript of phone call from Jacqueline Cochran to Nancy Love, July 3, 1944, in Cochran / Eisenhower.

5. Rickman, *Nancy Love and the WASP Ferry Pilots*, 194, as per the transcript.

6. Granger, *On Final*, 374.

7. *Thirty-Eight, In Memoriam.*

8. Memo, Stiles to Love, 17 July 44, in Nancy Love Collection donated by H. O. Malone, AF historian, TWU WASP Archives.

9. Weekly Summary, 9 August 1944, sent to Publications, attention Miss Wylie, in Love Collection from Malone, TWU.

10. Marx, 341.

11. Gott, *Women in Pursuit.* That figure is taken from the number of women pictured on 87–93 and listed on 96, 120, 140, 168, 198.

12. Barbara London, author interview October 6, 2006. Used also in *Nancy Love and the WASP Ferry Pilots*, by this author, 196.

13. Email from Iris Critchell to author dated Feb. 12, 2012.

14. Keil, *Those Wonderful Women*, 381, Cochran's Final Report on the Women Pilot Program.

15. Mrs. Love's comments on WASP available for transfer outside of Ferrying Division due to unsuitability on pursuit ships, dated June 5, 1944, in Jernigan/Archambault collection at TWU, BHG F3 #1, 206.

16. Esther Manning diary.

17. Granger, *On Final,* 396.

18. Memo signed by Betty Gillies, WASP Squadron Leader, 6 October 1944, at TWU, in the Jernigan/Archambault collection, 97.

19. LaFarge, "Women of the ATC,"148.

20. Letter from Ferrying Division commander (as of August 1, 1944) Gen. Robert E. Nowland, Subject: Inactivation of WASP addressed to Commanding General, ATC, and dated November 1, 1944.

21. Delphine Bohn qualified for her four-engine rating on June 8, 1944. Information found in her self-generated four-page biography. The

biography also notes that she received her pursuit rating on September 2, 1944. From the Women's Collection, WASP Archives, TWU.

22. Granger, *On Final*, A-29 / B.

23. Granger, *On Final*, Appendix B, A-17/B-A-66/B

24. Ruth Adams, Oral History. Interviewed 1989 by her 44-2 classmate Jean Hascall Cole, held at the WASP archives at TWU. Much of the information on Ruth throughout this chapter comes from her oral history and from the book by her classmate Jean Hascall Cole, *Women Pilots of World War II* (Salt Lake City: University of Utah Press, 1992).

25. Two sources: Gott, *Women in Pursuit*, 130; Turner, *Out of the Blue*, 370.

26. Romulus chart showing hours flown and ratings for every one of the ninety-two WASP ferry pilots stationed there. From IWASM, copy in author's hands.

Chapter 28: *The Fall Out*

1. Cochran's final report to General Arnold, TWU, WASP Archives.

2. http://en.wikipedia.org/wiki/William_H._Tunner, accessed May 23, 2012.

3. Marx, 340–341.

4. Marx, 351.

5. Marx, 342–345.

6. "Merry-Go-Round" was written by syndicated columnist Drew Pearson. His anti-WASP column first appeared August 6 in numerous newspapers. Accessed online July 28, 2012 at:

http://dspace.wrlc.org/doc/bitstream/2041/20601/b06f01-0806zdisplay.pdf#search="

7. Marx, 351, quoting letter from Major Geddings to Air Staff, Attention: Lt. Col. Rex Smith, August 11, 1944, Subject: Publicity on WASP in Ferrying Division.

8. Memorandum from Captain William R. Geddings, Public Relations Officer, to Nancy Love dated August 9, 1944, copy held at TWU.

9. The story is related in *The Originals*, by this author, 316–320.

10. Author interviews with Teresa James, June 1999 and March 2000.

11. LaFarge, Appendix II: Utilization of WASP by 552nd AAF Base Unit, 2nd Ferrying Group, ATC, Republic Field, Farmingdale, Long Island, NY, dated November 29, 1944.

12. Marx, 352-354. Associated Press dispatch of October 2, 1944, reported in the *Cincinnati Enquirer*, Oct. 3. And, War Department Bureau of Public Relations, Oct. 3, "AAF to inactivate WASP on December 20."

13. Marx, 355. The original TWX was received the morning of October 5, 1944. A copy of this TWX and of General Giles' handwritten note is in the WASP file at Ferrying Division Headquarters, Central Files.

14. Marx, 356. TWX sent the same day as the previous messages.

15. Marx, 358–359. Letter from Gen. Robert E. Nowland, Subject: Inactivation of WASP, addressed to Commanding General, ATC, and dated November 1, 1944.

Chapter 29: Ferrying—the Highs and the Lows

1. United States Air Force Museum, Aircraft Brochure featuring more than 175 aircraft of the U.S. Aircraft Museum, with aircraft photos, text, and specifications. Newly Revised Edition, 2003, 19.

2. Gott, *Women in Pursuit*, 65.

3. Delphine Bohn, "Catch a Shooting Star," Chapter 15, page 10 and separate manuscript, both held by TWU.

4. Dorothy Kocher Olsen, oral history TWU.

5. United State Air Froce Museum, Aircraft Brochure, 2003, 73.

6. Critchell, material from her memoir-in-progress, working title "76 Years With the Spirit of Flight."

7. Gott, *Women in Pursuit*, 190-191.

8. Gott, *Women in Pursuit,* 71.

9. Catherine Vail Bridge, oral history, TWU.

10. Rickman, *The Originals,* 305–306. Story related to author by Teresa James in a March 2000 interview.

11. Story told to author by Nancy Batson Crews and used in the author's book *The Originals*, 306.

12. Marie Muccie Genaro, oral history TWU.

13. Keil, *Those Wonderful Women,* 324.

14. Rickman, *Nancy Batson Crews*, 72.

15. Nancy Batson Crews interviews with the author, dated May 1999, and June, August, November and December 2000.

16. Jane Straughan, oral history, TWU.

17. Ruth Adams, oral history.

18. Vivian Cadman Eddy, oral history, September 9, 2005, in the TWU Archives, interviewer Sarah Rickman.

Part VIII—Denouement

1. *World War II Day by Day*, 576.

2. *World War II Day by Day*, 579, 587.

3. *World War II Day by Day*, 587.

Chapter 30: Twelve WASP Ferry Pilots Died

1. Virginia Luttrell Krahn, oral history, TWU.

2. Vivian Cadman Eddy, oral history.

3. Virginia Krahn, oral history.

4. *In Memoriam, Thirty-Eight American Women Pilots*, TWU.

Chapter 31: Facts of Life in the Ferry Command

1. Cochran's Comments, rebuttal to Capt. Walter Marx's original history, used as a source in this book. My specific copy of the source is found in the Jernigan/Archambault Collection, BHG F3 #1, beginning the bottom three lines on page 29 through the first full paragraph on page 30 (pages 25–26 of Cochran's Comments). The author has paraphrased Cochran's points to make them less wordy and easier to understand. Also found in LaFarge, "Women Pilots in the ATC," 132.

2. LaFarge, "Women Pilots in the ATC," 133.

3. Bohn, "Catch," Chapter 17, 5.

4. Helen McGilvery collection, TWU and copies in the author's hands.

5. Letter from Betty Gillies dated July 17, 1944, Helen McGilvery personal collection, copy at TWU and in the author's hands.

6. Using Byrd Granger's roster information in *On Final Approach.*

7. Marie Muccie Genaro, oral history TWU.

8. Turner, *Out of the Blue,* 184.

9. A WASP oral history, TWU.

10. Margaret Hardy Reeves Riviere, oral history, TWU.

11. Esther Manning Rathsfelder letters.

12. Inez Woodward Woods, oral history, TWU.

13. #55, "Women Pilots AAF," 105, fn 9, quoting a letter Lazarsky wrote to Jacqueline Cochran dated September 14, 1944.

14. #55, "Women Pilots AAF," 105, fn 9.

15. Virginia Lee Jowell Hagerstrom, oral history, TWU.

16. Gayle Bevis Reed, oral history, TWU.

17. Martha Wagenseil Davis Rupley transcript, Byrd Granger collection, TWU.

18. Jane Straughan, oral history, TWU.

19. Turner, *Out of the Blue,* 182.

20. Rickman, *The Originals,* 286-288. Initial source, multiple interviews with Teresa James, March 2000, and copies of papers from her collection in the author's hands.

21. Marx, 356-357. Letter 40-34 from AAF headquarters to commanding generals, all Air Forces, COs all AAF Bases, etc, Subject "Deactivation of WASP," 1 November 1944. This paragraph is paraphrased from Marx's interpretation.

22. LaFarge, Appendix II: Communication from AAF Headquarters, re: deactivation of WASP, dated November 1, 1944, 118-122; full text of AAF Letter # 40-34, Subject: Deactivation of WASP. To All Commanding Generals, AAF Policy Governing Separation of WASP.

23. LaFarge, Appendix II, 123-133, 168-178, contains letters from the bases queried.

24. LaFarge, Appendix II. The compilation includes all bases but Romulus. That base responded with a lengthy (19 pages) report found on pages 141-159.

25. Source www.hydraulicspneumatics.com, accessed July 13, 2014.

26. LaFarge, Appendix II: Letter from Lt. Joseph P. Tracy, Captain Air Corps Factory Control Officer, Curtiss-Wright Corporation, Buffalo, New York, dated 2 December 1944. Subject: Utilization of WASPs, 131-133.

27. Marx, 362-363, fn 26: Staff meeting notes, Ferrying Division Hq, December 15, 1944.

28. Marx, 363, fn 27: Weekly Summary of Activities, Ferrying Division Hq. Nov. 27-Dec. 2, 1944.

29. Marx, 363, fn 28: TWX.

30. Marx, 364-365, fn 31, Daily activity report, FD HQ, December 20, 1944.

31. Marx, 365, fn 32, Daily activity report, FD HQ, December 27, 1944.

32. Gill Robb Wilson, *New York Herald Tribune*, December 13, 1944.

33. Undated memo, Cochran to Arnold "Very personal and confidential," Cochran Collection, Eisenhower Library.

34. USAFA Oral History Interview #44, March 11-12, 1976; Recording of Jacqueline Cochran oral history interview, produced by Rohmann Joint Venture, USAFA, Yvonne Pateman Collection, 1976;

35. Phone interview by author with Bee Falk Haydu, WASP Class 44-7, June 6, 2006. This quote is also found in *Nancy Love and the WASP Ferry Pilots of WWII*, by the author, 230.

36. *World War II Day by Day*, 586.

Chapter 32: Women of the Ferry Command ... In Their Own Words

1. Transcript: An Oral History: Betty Huyler Gillies — Women's Auxiliary Ferrying Squadron. Interviewed by Dawn Letson, director, the Woman's Collection, Texas Woman's University (TWU), October 6, 1996, 21-22. On file in the WASP Archives, TWU.

2. http://www.absoluteastronomy.com/topics/San_Gorgonio_Pass. http://www.absoluteastronomy.com/topics/San_Gorgonio_Mountain.

http://www.absoluteastronomy.com/topics/San_Jacinto_Peak. Accessed Dec. 30, 2011.

3. Ruth Dailey Helm, oral history, TWU.

4. Gayle Bevis Ewing Reed oral history, TWU.

5. Jo Myers Wheelis, WASP class 43-5, oral history, interviewed by Dawn Letson

May 9, 2006, TWU.

6. Nancy Love had an A-20 at her disposal from late July to December 1944. She used it to make the rounds visiting her ferrying squadrons as needed. "It was a stripped-down G model that was sensitive on the controls, had a fuselage tank—good for long range—and a spirited personality all of its own." It was her favorite airplane. Nancy's first flight in the A-20 is listed in her logbook on July 22, 1944—from Washington to NCAAB (Wilmington) and back to Washington. Likely she picked it up that morning, flew down to Wilmington to show it off to Betty Gillies and the other girls, and returned that afternoon. Likely this is the flight Jo made with her, but no passenger is listed in the logbook. Author has a copy of the logbook.

7. Sylvia Dahmes Clayton oral history, TWU.

8. Anna Flynn Monkiewicz, oral history, TWU.

9. Monkiewicz, *My Piece of the Sky,* Chapter 18, Kindle version, taken from paragraphs 5-7.

10. Dorothy Kocher Olsen, oral history, TWU.

11. Karen Brewster, editor, *Boots, Bikes and Bombers: Adventures of Alaska Conservationist Ginny Hill Wood.* (Fairbanks, Alaska: University of Alaska Press, 2012.) 88-94. Biography of Ginny Hill Wood, WASP class 43-4.

12. Ruth Adams, oral history.

13. Article in *The Montclarion,* June 10, 1994: "Women's Airforce pilot reflects on World War II service" by Carolyn Younger. From the WASP Archives, TWU. Some information about Ruth Adams is taken from this article about her.

14. Mary C. "Jary" Johnson McKay, oral history TWU.

15. Iris Cummings Critchell, material from her memoir-in-progress, working title "76 Years With the Spirit of Flight."

16. Jean Landis oral history and information taken from the DVD *She Wore Silver Wings* featuring Jean. American Dream Cinema, © 2010, Spring Valley, California, AmericanDreamCinema.com.

Afterword

1. James Lea Cate and Wesley Frank Craven, eds., *The Army Air Forces in World War II, Vol. VI: Men and Planes* (Chicago: University of Chicago Press, 1955), 679-680.

2. Barbara London interviews.

3. Keil, *Those Wonderful Women,* 340–343.

Bibliography

Published Sources

Arnold, H.H. *Global Mission.* Blue Ridge Summit, PA: TAB Books, Harper & Row, 1949.

Bartels, Diane Ruth Armour. *Sharpie: The Life Story of Evelyn Sharp.* Lincoln, NE: Dageforde Publishing, 1996.

Brewster, Karen, ed. *Boots, Bikes, and Bombers: Adventures of Alaska Conservationist Ginny Hill Wood.* Fairbanks: University of Alaska Press, 2012.

Bunting, Edward, et al., eds. *World War II Day by Day.* London: Dorling Kindersley, 2001.

Cearley, George W., Jr. *Dallas Love Field (A Pictorial History of Airline Service).* Self Published, 1989.

Churchill, Jan. *From Delaware to Everywhere: New Castle Army Air Base, New Castle County Airport.* Dover, DE: Dover Litho Printing Co., 2007.

——. *On Wings to War: Teresa James, Aviator.* Manhattan, KS: Sunflower University Press, 1992.

Cochran, Jacqueline, and Maryann Bucknam Brinley. *Jackie Cochran: An Autobiography of the Greatest Woman Pilot in Aviation History.* New York: Bantam Books, 1987.

Cole, Jean Hascall. *Women Pilots of World War II.* Salt Lake City: University of Utah Press, 1992.

Curtis, Lettice. *The Forgotten Pilots: A Story of the Air Transport Auxiliary 1939–1945.* 4th ed. Cheltenham, England: Westward Digital Limited, 1998.

Douglas, Deborah G. *American Women and Flight Since 1940.* Lexington: University Press of Kentucky, 2004.

Fahey, James C. *U.S. Army Aircraft (Heavier Than Air) 1908-1946.* New York: Ships and Aircraft, 1946.

Gott, Kay. *Women in Pursuit.* Self Published, 1993.

——. *Women's Flying Training Detachment, Women Pilots Trained at Houston, Texas: A Tribute to Lt. Alfred Fleishman.* Self Published, 1992.

Granger, Byrd Howell. *On Final Approach: The Women Airforce Service Pilots of W.W.II.* Scottsdale, AZ: Falconer Publishing Company, 1991.

Graves, Celeste. *A View from the Doghouse of the 319th AAFWFTD.* Bloomington, IN: AuthorHouse, 2004.

Hodgson, Marion Stegeman. *Winning My Wings.* Albany, TX: Bright Sky Press, 2005.

Holm, Jeanne. *Women in the Military: An Unfinished Revolution.* Rev. ed. Novato, CA: Presidio Press, 1992.

Keil, Sally Van Wagenen. *Those Wonderful Women in Their Flying Machines: The Unknown Heroines of World War II.* Rev. ed. New York: Four Directions Press, 1990.

LaFarge, Oliver (Lt. Colonel). *The Eagle in the Egg.* Boston, MA: Houghton, Mifflin Co., the Riverside Press, 1945.

Makanna, Philip, and Jeffrey Ethell: *Ghosts: Vintage Aircraft of World War II.* Charlottesville, VA: Thomasson-Grant, 1987.

Mason, Daniel W. *Detroit Metro Airport: The Images of Aviation.* Charleston, SC: Arcadia Publishing, 2011.

Matz, Onas P. *History of the 2nd Ferrying Group, Ferrying Division, Air Transport Command.* Seattle, WA: Modet Enterprises, Inc., 1993.

Monkiewicz, Anna Flynn. *My Piece of the Sky.* Amazon Digital Services, Inc., April 4, 2013. Kindle.

Rich, Doris L. *Jackie Cochran: Pilot in the Fastest Lane.* Gainesville: University Press of Florida, 2007.

Rickman, Sarah Byrn. *Nancy Batson Crews: Alabama's First Lady of Flight.* Tuscaloosa: The University of Alabama Press, 2009.

——. *Nancy Love and the WASP Ferry Pilots of World War II.* Denton: The University of North Texas Press, 2008.

——. *The Originals: The Women's Auxiliary Ferrying Squadron of World War II.* Sarasota, FL: Disc-Us Books, 2001.

Scharr, Adela Riek. *Sisters in the Sky.* 2 vols. St. Louis: Patrice Press, 1986.

Serling, Robert. *Eagle: The Story of American Airlines.* New York: St. Martin's/ Marek, 1985.

Seymour, Dawn, Clarice I. Bergemann, Jeannette J. Jenkins, and Mary Ellen Keil. *In Memoriam: Thirty-eight American Women Pilots.* Denton: Texas Woman's University Press, ND.

Simbeck, Rob. *Daughter of the Air: The Brief Soaring Life of Cornelia Fort.* New York: Atlantic Monthly Press, 1999.

The Ninety-Nines, Yesterday—Today—Tomorrow. Paducah, KY: Turner Publishing Company, 1996.

Tillman, Barrett. *Forgotten Fifteenth: The Daring Airmen Who Crippled Hitler's War Machine.* Washington D.C.: Regnery History/Regnery Publishing, 2014.

Tunner, William H. (Lt. General) with Booton Herndon. *Over the Hump.* New York: Duell, Sloan and Pearce, 1964.

Turner, Betty Stagg, *Out of the Blue and Into History.* Arlington Heights, IL: Aviatrix Publishing, 2001.

United States Air Force Museum, Aircraft Brochure featuring more than 175 aircraft of The Museum of the United States Air Force, with aircraft photos, text, and specifications. Newly Revised Edition, 2003.

Verges, Marianne. *On Silver Wings.* New York: Ballantine Books, 1991.

Government Historical Studies, WWII

"History of the Air Transport Command, Women Pilots in the Air Transport Command," prepared by the Historical Branch, Intelligence and Security Division, Headquarters, Air Transport Command in accordance with ATC Regulation 20-20, AAF Regulation 20-8, and AR 345-105, as amended. Author is Lt. Col. Oliver LaFarge, official historian for the Air Transport Command. This is the accepted history of the women ferry pilots of the ATC. Referred to in Endnotes as LaFarge, "Women Pilots ATC." Also Appendix II of that document.

IRIS # 145345. WASP Activities in the Air Training Command Volume 3. Air Force Historical Research Agency, Department of the Air Force, Maxwell AFB, Alabama.

"Women Pilots in the Ferrying Division, Air Transport Command." A history written in accordance with AAF Regulation No. 20-8 and AAF Letter 40-34; unpublished. Capt. Walter J. Marx is the author. Held in the Nancy Love Private Collection. A copy also is in the author's hands. Referred to in Endnotes as Marx, "Women Pilots Ferrying Division."

"Women Pilots AAF, 1941–1944." *Army Air Forces Historical Studies*: No. 55. March 1946. This document is part of the WASP Archival Collection, Texas Woman's University Library, Denton, Texas. It is also part of the Jacqueline Cochran Collection, Dwight D. Eisenhower Presidential Library, Abilene, Kansas. Referred to in Endnotes as #55, "Women Pilots AAF."

Unpublished Sources

Bohn, Delphine. "Catch a Shooting Star," copies in hands of the author and the WASP Archives, Texas Woman's University, Denton.

Critchell, Iris Cummings. Material from her memoir-in-progress, working title "76 Years With the Spirit of Flight."

DVDs

She Wore Silver Wings, featuring Jean Landis, WASP Class 43-4. Spring Valley, California, American Dream Cinema, 2010

Wings of Silver: The Vi Cowden Story, featuring Vi Thurn Cowden WASP Class 43-4, a 33-minute DVD. California: The Monkey House, 2010.

Articles

Crane, Mardo. "The Women With Silver Wings," Part One, *The 99 News.* Special Issue, 1978.

Carter, Rowland. "The Ladies Join the Air Forces." *Flying Magazine,* December 1942.

Douglas, Deborah G. "WASPs of WAR." *Aviation Heritage.* January 1999.

Nyman, Geri Lamphere. "Bean Salads and Bamboo Bombers." *WASP News.* Summer 2011.

Rickman, Sarah Byrn. "Barbara Erickson: From 'Rosie the Riveter' to B-17 Pilot." *Air Power History.* Summer 2005.

——. "Born to Fly," *Kansas Heritage*,, Winter 2006. Article about Betsy Ferguson, WAFS.

—— and Barbara Erickson London. "The Lady Flew the B-17." *Aloft: The Museum of Flight Magazine* (Seattle). September/October 2005.

——. "Women With Wings: The WASP in Michigan." *Michigan History.* July/August 2010.

WASP Oral Histories

WAFS

Author's personal conversations with: Nancy Batson, Barbara Erickson, Barbara Donahue, Teresa James, Gertrude Meserve, Florene Miller, and Barbara Poole. Phone conversations with: Phyllis Burchfield. Mail correspondence with Bernice Batten.

Nancy Batson Crews audiotapes done by author and in author's possession: June, August, November, and December, 2000.

Phyllis Burchfield Fulton, telephone interview with author, December 2005.

Betty Huyler Gillies, oral history done by Dawn Letson, TWU, September 1996.

Teresa James, author interviews with James, June 1999 on video (TWU has copy) and audio March 2000 (author has tape).

Gertrude Meserve Tubbs LeValley, oral history done by author for TWU December 2004 (audio) and 1999 (video), TWU has copy.

Barbara Erickson London, oral history done by the author for TWU March 17 and 19, 2004, plus personal conversations over several years.

WASP

Iris Cummings Critchell: Author's personal conversations with Iris, over several years.

Unless otherwise noted, the following oral histories are held by Texas Woman's University, Denton: the WASP Collection and the WASP Archives.

WAFS

Betty Huyler Gillies, October 6, 1996.

Gertrude Meserve Tubbs LeValley, December 2004

Barbara Erickson London, March 17 and 19, 2004.

Class 43-1

Betty Tackaberry Blake, August 6, 2003.

Mary Lou Colbert Neale, October 4, 2002

Geri Lamphere Nyman, October 4, 1996.

Jane Sincell Straughan, October 4, 1996.

Class 43-2

Catherine "Cappy" Vail Bridge, October 27, 2003.

Marion Schorr Betzler Brown, May 27, 1998.

Ruth Dailey Helm, October 3, 2002.

Marie Muccie Genaro, July 12, 2002.

Mary Catherine "Jary" Johnson McKay, October 14, 1994.

Martha Wagenseil Davis Rupley, circa 1980s done by Martha herself, transcript in Byrd Granger collection.

Elizabeth Whitlow Rummerfield Smith, August 4, 1999.

Margaret Ann Hamilton Tunner, May 8, 1998.

Class 43-3

Lois Brooks Hailey, May 21, 1994.

Lillian Connor Roberts-Risdon, October 7-8, 1996

Lois Hollingsworth Ziler, May 21, 1994

Class 43-4

Virginia Lee Jowell Hagerstrom, October 13, 1994.

Virginia Luttrell Krahn, February 9, 1997

Jean Landis, May 13 and September 16, 2012

Dorothy Kocher Olsen, May 2005.

Margaret Hardy Reeves Riviere, March 2006.

Alice Jean "A.J." May Starr, February 27, 2011.

Inez Woodward Woods, September 2005 for TWU. Also, by Ruth Florey for the Commemorative Air Force Library, Midland-Odessa, Texas, no date given.

Class 43-5

Sylvia Dahmes Clayton, October 5, 2002.

Vivian Cadman Eddy, September 2005.

Gayle Bevis Ewing Reed, October 2, 2004.

Jo Myers Wheelis, May 9, 2006.

Class 43-6

Adeline Wolak Ellison, September 30, 2004.

Anna Flynn Monkiewicz, October 2006.

Class 44-2

Ruth Adams, circa 1989 by her classmate, Jean Hascall Cole.

Class 44-6

Elinor Fairchild Stebbins, May 18, 2006.

Other Oral Histories held by TWU

Leoti "Dedie" Deaton, WASP Chief Establishment Officer, interviewed by Ziggy Hunter, March 18, 1975.

Leoti "Dedie" Deaton, WASP Chief Establishment Officer, interviewed by Doris Tanner, March 20, 1982.

Eduardo Escallon, nephew of WAFS Helen Schmidt McGilvery, interviewed by Sarah Byrn Rickman, February 2002.

Honey Fulton Parker, sister of WAFS Dorothy Fulton, interviewed by Sarah Byrn Rickman, September 2004.

Other Oral Histories

Jacqueline Cochran, Director of Women Pilots

- USAFA Oral History Interview #44, March 11-12, 1976; Recording of Jacqueline Cochran oral history interview, produced by Rohmann Joint Venture, USAFA, Yvonne Pateman Collection, 1976;
- The Reminiscences of Jacqueline Cochran (1960), pp. 56-57, in the Columbia (University) Center for Oral History Collection (hereafter referenced as CCOHO). Photocopy in author's hands.

Diaries

Teresa James, September 1942 to February 1943, from Teresa James's personal collection. A copy is in author's hands and in WASP Archives, Texas Woman's University, Denton.

Esther Manning Rathsfelder Shively Westervelt, 1944, courtesy of her daughter Julie Shively.

Collections

Archambault/Jernigan Collection, Texas Woman's University, WASP Archives, Denton, Texas.

Jacqueline Cochran Collection held at the Dwight D. Eisenhower Library, Atchison, Kansas.

Betty Huyler Gillies Collection, the International Women's Air and Space Museum, Cleveland, Ohio.

Byrd Howell Granger Collection, TWU WASP Archives.

Nancy Harkness Love private collection, held by her daughter, Margaret Love.

Nancy Love collection, held by author.

Nancy Love Collection, TWU WASP Archives.

H.O. Malone Collection, TWU WASP Archives.

Helen McGilvery personal papers, TWU WASP Archives, author has copies.

Helen Richards personal papers, TWU WASP Archives, author has copies.

Dorothy Scott Collection, Letters Home, TWU WASP Archives, author has copies and has written Scott's biography based on them: *Finding Dorothy Scott: Letters of a Womens Airforce Service Pilot*, forthcoming from Texas Tech University Press, June 2016.

Index